THE
DISCHARGE OF ELECTRICITY THROUGH GASES

LECTURES DELIVERED ON THE OCCASION OF
THE SESQUICENTENNIAL CELEBRATION
OF PRINCETON UNIVERSITY

BY

J. J. THOMSON

PROFESSOR IN THE UNIVERSITY OF CAMBRIDGE

WITH DIAGRAMS

Copyright © 2018 Read Books Ltd.
This book is copyright and may not be
reproduced or copied in any way without
the express permission of the publisher in writing

British Library Cataloguing-in-Publication Data
A catalogue record for this book is available from
the British Library

THIS BOOK IS DEDICATED

TO

THE MEMBERS OF THE CLASS WHO ATTENDED
THE AUTHOR'S LECTURES AT
PRINCETON,

WHOSE SYMPATHY AND KINDNESS HE CAN
NEVER FORGET.

THE DISCHARGE OF ELECTRICITY
THROUGH GASES

PREFACE

THE following pages contain an expansion of four lectures on Discharge of Electricity through Gases, given at the University of Princeton, New Jersey, in October, 1896. In the hope of making the work more useful, I have added some results which have been published between the delivery and the printing of the lectures.

J. J. THOMSON.

CAMBRIDGE, August, 1897.

THE
DISCHARGE OF ELECTRICITY
THROUGH GASES

CONTENTS

	PAGE
THE DISCHARGE OF ELECTRICITY THROUGH GASES	3
Communication of a Charge of Electricity to a Gas	5
Electrification of Gas by Chemical Means	7
Electrification of Gases liberated by Electrolysis	10
Formation of a Cloud round Electrified Gas	11
Electrification by the Splashing of Liquids	17
Electrification of a Gas by the Aid of Röntgen Rays	29
Uranium Radiation	56
PHOTO-ELECTRIC EFFECTS	61
Electrification of Gases by Glowing Metals	81
Electrification in the Neighbourhood of an Arc Discharge	86
The Arc in Hydrogen	89
Conduction through Hot Gases	97
Conduction by Flames	101
Effect of a Discharge in making a Gas a Conductor	105
Electrolysis in Gases	121
Short Sparks	129
Medium Sparks	130
Long Sparks	130
Transport of one Gas through Another	134

CONTENTS

	PAGE
CATHODE RAYS	137
Properties of the Cathode Rays	139
Thermal Effects produced by the Rays	145
Mechanical Effects produced by the Cathode Rays	146
Action of a Magnet upon the Cathode Rays	150
Paths of the Rays in Different Gases	155
Shape of the Path of the Rays	161
Electric Charge carried by the Cathode Rays	161
Repulsion of Cathodic Streams	169
Diffuse Reflection of Cathode Rays	179
Transmission of Cathode Rays	181
Lenard's Experiments	182
Magnetic Deflection of the Rays	188
Theories of the Nature of the Cathode Rays	189

THE
DISCHARGE OF ELECTRICITY
THROUGH GASES

THE variety and complexity of the electrical phenomena which occur when matter is present in the electric field are in marked contrast to the simplicity of the phenomena when the ether alone is involved. The latter, as far as our knowledge of them extends, are fully explained by laws which can be expressed mathematically by six very simple differential equations. Indeed, the phenomena in the ether appear to be even simpler in character than we could expect *a priori*, for the equations we have alluded to as covering all the phenomena which have been observed are true only when the ether is at rest. The agreement of theory and experiment justifies the tacit assumption involved in their use, that the ether remains at rest even when exposed to considerable mechanical forces, as it is when an electric wave is passing through it. The negative result of experiments made to detect the motion of the ether in the electro-magnetic field also tends to justify this assumption.

When, however, we consider electric phenomena in which matter plays a part, we find qualities coming into prominence which hardly appear at all as long as we confine our attention to the ether; thus the idea of a charge of electricity, which is perhaps in many classes of phenomena the most prominent idea of all, need not arise, and in fact does not arise, as long as we deal with the ether alone.

The questions which occur when we consider the relation between matter and the electric charge carried by it — such as, the state of the matter when carrying this charge, the effect produced on this state when the sign of the charge is changed — are some of the most important in the whole range of Physics. The close connection which exists between electrical and chemical phenomena — as shown, for example, in electrolysis — indicates that a knowledge of the relation between matter and electricity would lead not merely to an increase of our knowledge of electricity, but also of that of chemical action, and might indeed lead to an extension of the domain of electricity over that of chemistry.

If we wish to study the relation between matter and electricity, the most promising course is to begin with the relation between electricity and matter in the gaseous state; for the properties of a gas and the laws it obeys are simpler than for either a solid or a liquid, it is the state of matter which has been most studied, while the Kinetic Theory of Gases supplies us with the means of forming a mental picture of the processes going on in a gas which is lacking for matter in its other states.

Communication of a Charge of Electricity to a Gas.

One of the most striking phenomena connected with the electrical properties of gases is the difficulty of directly communicating a charge of electricity to a gas in its normal condition. A very simple instance will suffice to show this: let us take the case of a charged metal plate which is insulated so well that there is no leakage of electricity across its supports ; let this plate be in contact with air or any other gas at a moderate temperature, and let it be screened off from ultra-violet light and Röntgen rays, — then the evidence of the best experiments we have, proves that under these conditions the plate will suffer absolutely no loss of charge, provided the surface density of its electrification is less than a certain value. Thus, though myriads of molecules of the gas strike against the charged surface, they rebound from it without any electrification. To fully appreciate the significance of this result, we must remember the very large charges that can be carried by the gas under other conditions. The phenomena of electrolysis show that the charge on each unit of surface of the plate could be carried thousands of millions times over by a cubic centimetre of hydrogen at normal pressure and temperature. We must, I think, conclude that the inability of a gas (which when in a certain state has such an enormous capacity for carrying electricity) to take up when in its normal condition any of the charge of electricity from a body against which it strikes is very significant and suggestive.

Another fact which exhibits in perhaps even a more striking way the inability of the molecules of a gas to take up an electric charge is that the vapour arising from an electrified liquid is quite free from any charge of electricity. We owe this discovery to two American men of science; for in 1761, Kinnersley of Philadelphia, in a letter to Franklin, stated that he found the steam arising from electrified water was not itself electrified. This result seems to have been overlooked for a long time; and in 1883 another American physicist, Blake, made a very complete investigation of the subject,[1] and found that the vapour arising from boiling mercury was not electrified, however strongly the mercury itself might be. Blake's results have

[1] Wiedemann's Annalen, 19, p. 518, 1883.

been confirmed by Sohncke,[1] and quite recently by Schwalbe.[2] In these experiments the molecules of the unelectrified vapour came from or through an electrified surface, and the conditions here seem such that if a molecule could ever get electrified by contact, it would do so in these experiments. Thus we see that when an electrified liquid evaporates, the electrified particles are left behind, just as the salt in a salt solution is left behind on evaporation.

Electrification of Gas by Chemical Means.

Chemical action is so frequently attended by electrical separation that we might expect that the most likely way to electrify a gas would be to make it one of the parties in a chemical reaction. Examples of electrification, apparently in a gas, produced by this method have been known for a long time, though it is only in recent years that experiments have been made to show that the electrification in these cases is not, in all cases, carried by dust which may either be present in the gas to begin with, or produced by the chemical reaction itself.

One of the earliest-known cases of electrification produced in a gas by chemical action is that of the combustion of carbon. Pouillet[3] found that when

[1] Wiedemann's Annalen, 34, p. 925, 1888.
[2] Ibid., 58, p. 500, 1896.
[3] Pogg. Ann., ii. 422.

a carbon cylinder was being burnt, the cylinder was negatively electrified, while there was positive electrification in the gas over the cylinder. Lavoisier and Laplace[1] showed that the same effect takes place with glowing coal. Reiss[2] proved that there was positive electricity in the air near a glowing platinum spiral. Pouillet[3] found that when a jet of hydrogen was burnt in air there was negative electrification in the unburnt hydrogen in the jet.

Another case of electrification produced in a gas by chemical action was discovered by Lavoisier and Laplace,[4] who found that when hydrogen is rapidly liberated by the action of sulphuric acid on iron, a strong positive electrification comes off. This, with other cases of electrification by chemical action, was investigated by Mr. Enright.[5] As a good deal of spray is produced by the bubbling of the hydrogen through the sulphuric acid, it has been suggested that the spray and not the gas may in this case be the carrier of the electrification. To test this point, Mr. Townsend recently made in the Cavendish Laboratory a series of experiments on the electrification produced when a gas is liberated by chemical action. He found that

[1] Phil. Trans., 1782.
[2] Reiss, Reibungselektricität, vol. i. p. 267.
[3] Pogg. Ann., ii. 426.
[4] Mémoires de l'Acad. des Sciences, 1782.
[5] Phil. Mag. [5], 29, p. 56, 1890.

when the hydrogen produced by the action of strong sulphuric acid on iron or zinc was passed through tubes fitted with plugs of tightly packed glass wool it retained after its passage through these plugs a strong positive electrification, thus showing that no ordinary spray could be the carrier of the electrification.

Using an open beaker, Mr. Townsend[1] found that when the mixture of sulphuric acid and iron was heated to about 94° C., there was at first strong positive electrification at the mouth of the beaker when the chemical action was very vigorous and the gas was coming off with great rapidity. But as the temperature fell and the rate of evolution of hydrogen diminished, the positive electrification diminished and finally changed to negative. It was found, however, that the *negative* electrification, unlike the positive, was completely stopped not only by a plug of glass wool, but even by a layer of wire gauze. This seems to indicate that the negative electrification is carried by coarse spray, while the positive is on the hydrogen, or at any rate on much smaller carriers. If there is positive electrification in the hydrogen, there must be an equal quantity of negative in the sulphuric acid; if this is dashed into spray, the spray will be negatively charged. The experiments indicate that

[1] Proceedings of Cambridge Philosophical Society, 1897.

when the gas is not coming off very vigorously, the negative electrification carried from the vessel exceeds the positive carried off by the gas; while when the gas comes off with great rapidity, the positive electrification carried by it far exceeds the negative carried by the spray.

Mr. Townsend investigated several other cases of electrification produced when a gas is liberated by chemical action. He found that when chlorine was liberated by the action of hydrochloric acid on manganese dioxide, the chlorine had a strong positive electrification, and that when potassium permanganate was heated, there was positive electrification in the oxygen evolved.

Electrification of Gases liberated by Electrolysis.

Mr. Townsend has also found that when a strong current is sent through a solution of sulphuric acid, so that there is a copious liberation of hydrogen at one terminal and of oxygen at the other, there is positive electrification in the hydrogen, while the oxygen is either apparently unelectrified or has a very small positive charge. It is perhaps to the point to mention that in this case the oxygen is liberated by a secondary process; the negative ion is SO_4, and the oxygen is liberated by the chemical action of this ion on the water. The positive electrification in the hydrogen is very much influenced by temperature. At the ordinary temperature of the Laboratory there is very little electrification; when, however, the temperature is raised to 40–50°C., the electrification is very strong.

When a solution of caustic potash is electrolyzed, there is, on the other hand, very little electrification in the hydrogen, while the oxygen is negatively electrified, though the amount of this electrification is not nearly so large as that on the hydrogen in the preceding experiment. In this case the hydrogen is liberated by secondary chemical action, and the great diminution in its electrification as compared with the previous case seems to show that a gas is more likely to be electrified from electrolysis when it forms one of the ions than when it is liberated by secondary chemical action. The amount of electrification in the oxygen rapidly increases with the temperature. If oil is added to the caustic potash solution, the sign of the electrification in the oxygen changes, and the oxygen is positively instead of negatively electrified. The nature of the electrodes has a considerable influence on the amount of electrification which comes off with the gas.

Formation of a Cloud round Electrified Gas.

Mr. Townsend has discovered that electrified gas possesses the remarkable property of producing a fog when admitted into a vessel containing aqueous vapour. This fog is produced even though the vessel is far from saturated with moisture and does not require any lowering of temperature such as would be produced by the sudden expansion of the gas in the vessel in which the fog is produced.

The method used by Mr. Townsend was to send the electrified gas, produced most conveniently by the rapid electrolysis of a solution of sulphuric acid or caustic potash, according as the gas was to be positively or negatively electrified, through a series of tubes, some of which were tightly packed with glass wool, whilst others contained strong sulphuric acid through which the gas bubbled. The electrified gas finally passed into the atmosphere, where it formed a dense cloud which slowly settled down.

We can, from observations made on this cloud, calculate the charge carried by each of the electrified particles of the gas. For by placing the beaker in which the cloud is formed inside an insulated metal vessel of known capacity connected with an electrometer, we can calculate, from the deflection of the electrometer when the electrified gas goes into the beaker, what is the total charge of electricity in the cloud. Thus, if we know the number of water particles in the cloud, we can calculate the charge of electricity on each. We can get the number of particles in the cloud in the following way. The rate at which the cloud falls gives us the radius and therefore the weight of each drop, since

$$a^2 = 4.5 \frac{\mu v}{a},$$

where v is the velocity with which a drop falls, a the radius of the drop, μ the coefficient of viscosity of the gas through which the gas falls.

The weight of the whole collection of drops forming the cloud can be determined by weighing the beaker with the cloud in, then blowing out the cloud and weighing again. Dividing the weight of the cloud by the weight of a drop, we get the number of drops in the cloud, and then dividing the charge of electricity on the whole cloud by the number of drops, we get the charge carried by each drop.

The results of a series of measurements made by Mr. Townsend were that the radius of the drop in the cloud formed by negatively electrified oxygen was 8.1×10^{-5} cm., the charge carried by it 3.1×10^{-10} electrostatic units, the size of the drop formed by positively electrified oxygen was 6.8×10^{-5} cm., and the charge carried by it, 2.8×10^{-10} electrostatic units. The difficulty of the measurement prevents us from attaching importance to the slight difference between the positive and negative charges.

The charge on the drop formed by positively electrified hydrogen was about half that of oxygen.

The charges carried by the electrified particles of oxygen and hydrogen agree within the limits of errors of experiment with those deduced from the electro-chemical equivalents of these substances. The calculation of the atomic charge from the electro-chemical equivalent involves a knowledge of the number of molecules in a cubic centimetre of a gas at standard temperature and pressure, and all that we at present know about this number is that it lies between the limits 10^{18} and 10^{21} (Boltzmann, Vorlesungen über Theorie der Gase); thus all that we can infer about the atomic charge from the values of the electro-chemical equivalent is that it is between 10^{-8} and $\times 10^{-11}$.

It is probable that in the present state of our knowledge as to the mass of a particle of oxygen or hydrogen, the atomic charge can be determined more accurately from observations on the cloud formed by the electrified particles than from the electro-chemical equivalent. In calculating the charge carried by each particle from the weight and rate of fall of the cloud, we have assumed that each drop in the cloud is associated with one and only one charged particle.

In a book on the Applications of Dynamics to Physics and Chemistry (p. 164) I have shown that the presence of an electric charge on a drop of water tends to prevent evaporation, and will when the drop is very small neutralize the effect of surface tension, which tends to promote evaporation. It is only, however, for exceedingly small drops that the effects of electrification balance those of surface tension; the size of the drop, when these effects just neutralize each other, is given by the equation,

$$a^3 = \frac{e^2}{16\,\pi T},$$

where a is the radius of the drop, e the charge of electricity on it, and T the surface tension. When the radius is greater than the value given by this equation the drop evaporates more readily than a plane surface. With the charge deduced from the preceding experiments the limiting size of the drop would be only 10^{-8}, whereas the rate of fall shows that the radius of the drop is in reality about 8×10^{-5}. This seems to indicate that the drops contain something besides pure water.

The appearance of the cloud and the size of its particles depends upon the sign of the electrification; thus the particles in the cloud formed by negatively electrified oxygen are larger than those formed by positively electrified oxygen. This would seem to indicate that a positively electrified drop of water evaporated more rapidly than a negatively charged one of the same size. Few direct experiments on the evaporation of electrified water surfaces seem to have been made. Mr. Crookes,[1] from some experiments he made on this point, came to the conclusion that a *negatively* electrified surface of water evaporated more rapidly than an unelectrified one. Mascart [2] came to the conclusion that an electrified surface, whether the electrification was positive or negative, evaporated more rapidly than an unelectrified one; while Wirtz [3] found that electrification diminished the rate of evaporation of dust-free water, and that positive electrification had more effect than negative.

Even when there is no cloud to be seen, there is evidence that an electrified particle of gas is the centre of an aggregate of some kind which is very large compared with the dimensions of a molecule. This is shown by an experiment made by Mr. Townsend, where dust-free electrified hydrogen was put into a porous pot. The rate at which hydrogen in its normal condition escaped from the pot was determined, and then by connecting the porous pot with an inductor connected with an electrometer the rate at which the charge escaped from the pot was determined: if the electrified

[1] Crookes, Proc. Roy. Soc., 50, p. 88, 1891.
[2] Mascart, Comptes Rendus, 86, p. 575, 1878.
[3] Wirtz, Wied. Ann., 37, p. 516, 1889.

particles had been simple molecules, the rate at which the charge escaped would have been equal to the rate of escape of the hydrogen, whereas the experiment showed that in reality the rate at which the charge escaped was only a small fraction of the rate at which the hydrogen escaped. This was not due to the electrified gas sticking in the pores of the pot, for if the gas were afterwards blown out of the pot, all but a fraction of the charge went out with it.

Electrification by the Splashing of Liquids.

One of the most effectual ways of charging a gas with electricity is by means of the splashing of liquids. It had been known for a long time that there was something anomalous in the condition of atmospheric electricity at the feet of waterfalls, where the water fell upon rocks and broke into spray. Lenard[1] investigated the subject with great thoroughness, and found that when a drop of water splashed against a metal plate, a positive charge went to the water, while there was negative electrification in the surrounding air. This electrical separation is even more marked in the case of mercury than in that of water. A very simple way of showing the effect is to shake some mercury up vigorously in a bottle and then draw off

[1] Lenard, Wied. Ann., 46, p. 584, 1892.

the air. This will be found to have a negative charge; this charge is carried neither by dust nor by spray, for it will remain after the air has been sucked through glass wool, or even through a coarse porous plate. Lord Kelvin[1] has shown that the reciprocal process of bubbling air through water also gives rise to electrification, the air which has bubbled through the water being negatively electrified.

To investigate the laws of electrification produced by splashing, it is often more convenient to measure the positive charge on the drop rather than the negative charge in the air. This can conveniently be done by the arrangement represented in Fig. 1. A known quantity of the liquid to be investigated is allowed to fall through the funnel A and falls on the metal plate and saucer C, which is carefully insulated and connected with one pair of quadrants of an electrometer; a blower sends a strong current of air over the plate and blows away the air from the neighbourhood of the plate, so as to prevent the negative electrification in the air interfering with the indications of the electrometer. The deflection of the electrometer when a given quantity of liquid falls on the plate may be taken as an indication of the electrification produced by the splashing of the

[1] Kelvin, Proc. Roy. Soc., 47, p. 335, 1894.

DISCHARGE OF ELECTRICITY THROUGH GASES 19

liquid. By the aid of an instrument of this kind we can investigate the circumstances which affect the electrification. This electrification is influ-

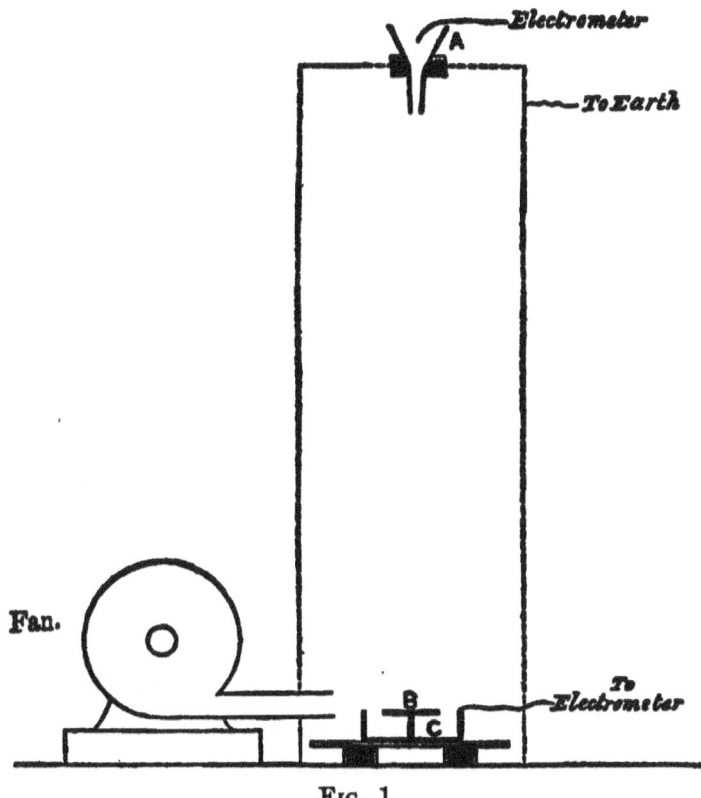

Fig. 1.

enced to a remarkable extent by minute changes in the composition of the liquid. Lenard found that the electrification in the air in the neighbourhood of the splashing place, which was very marked with the exceptionally pure water of Heidelberg, was almost insensible with the less pure water of Bonn; while the splashing of a weak salt solution electrified the air in the neighbourhood positively instead of negatively as when the water was pure. Thus, while the splashing of rain electrifies the air negatively, the breaking of waves on the seashore electrifies it positively.

In some experiments which I made on this subject, I found that the effects produced by exceedingly minute traces of some substances were most surprisingly large. Rosaniline, for example, is a substance which has very great colouring power, so much so, indeed, that the colour imparted to a large volume of water by a small quantity of rosaniline is sometimes given as an instance of the extent to which matter can be subdivided; yet I found that the change produced in the electrification, due to the splashing of drops, was appreciable in a solution so weak as to show no trace of colour. The effect of fluorescent solutions on the electrification of drops is especially great, but different kinds of solution act in different ways. Thus rosaniline and methyl violet reverse the effect, — that is, they make the electrification on the drop negative in the air; while eosine and fluorescein, on the other hand, increase the normal effect, — that is, they make the drop more strongly electrified positively than a drop of pure water, and produce a greater negative electrification in the air.

The electrical effects of weak solutions are much greater than those of strong ones; in fact, strong solutions of all substances tested gave little or no electrical effects by splashing. The effect produced by the addition of foreign substances to water may be represented by curves in which the abscissæ are proportional to the amount of the substance added to the water, and the ordinates to the electrification produced by splashing. We find that these curves are of three types, a, β, γ. Fig. 2.

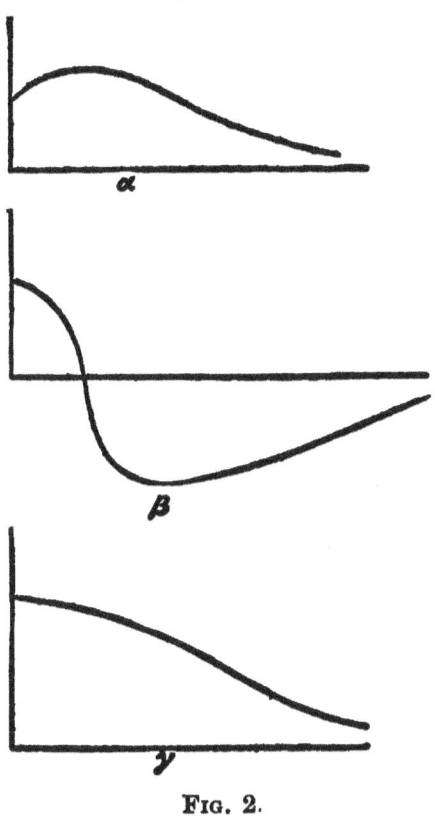

FIG. 2.

Curves of the type a represent the behaviour of solutions of phenol, eosine, fluorescein, where the addition of a small quantity of the substance increases the electrical effect.

Curves of the type β represent the behaviour of solutions of potassium permanganate, chromium binoxide, hydrogen peroxide, rosaniline, and methyl violet; here the addition of the substance begins by diminishing the electrification and finally reverses it.

Curves of the type γ represent the behaviour of solutions of zinc chloride, hydrochloric and hydriodic acids, and, in fact, of most inorganic salts and acids; here the addition of the substance produces a diminution but not a reversal of the electrification.

The addition of strong oxidizing agents to the water seems, on the whole, to tend to reverse the normal effect,— that is, it tends to make the electrification in the air positive, — while the addition of reducing agents seems to increase the normal effect.

The amount and even the sign of the electrification depends on the gas through which the drops fall; thus I found that when the drops fell through steam no electrification was produced; while when they fell through hydrogen which had been very carefully freed from any trace of air, the electrification in the hydrogen was very small but positive; that is, of the opposite sign to the electrification in air.

There does not appear to be any appreciable separation of the positive and negative electrification until the drop splashes against the plate; thus a drop of rain falling through the air does not appear to leave a trail of negative electricity behind it.

We can explain the electrification produced by the splashing of drops if we suppose that at the surface of each drop we have a double electrical layer; that is, a spherical layer of electrification of one kind surrounded by a closely fitting concentric layer of electricity of the opposite sign, the charge of positive electricity on one layer being equal to the charge of negative electricity on the other. In the case of pure water the positively electrified layer is next the water, the negatively electrified one next the air.

When the water-drop strikes against the plate, this double layer gets very roughly treated whilst the great increase which takes place in the area of the drop is in progress. The double layer gets torn asunder, the negative coating staying in the air, the positive one on the drop.

The suggestion that the electrification due to the splashing of drops is produced by the rupture of an electrical double layer is due to Lenard; and though it is perhaps somewhat surprising that the disturbance produced by splashing should produce so great a separation of the coats of the layer, the existence of the layer is proved in many cases by the definite contact difference of potential which occurs at the junction of two dissimilar substances. Though this contact difference of potential has only been measured when one of the substances is a conductor, we must, if we accept the electrification due to the splashing of drops as evidence of the existence of this double layer, suppose that the double layer is almost invariably present at the junction of different substances. We find, for example, that electrification is produced by the splashing of drops of substances as unlike as paraffin-oil, water, and mercury.

The electrification produced by splashing shows that the coatings of the layer can be separated by mechanical means; but if this is the case, and all substances are covered with an electrical double layer, the production of electrification by rubbing is very simply explained.

The existence of an electric double layer at the surface of separation of two bodies implies the existence at the surface of the body of a layer of matter which is neither quite of the nature of one substance or of the other; it thus implies a certain amount of chemical combination, or rather the first stages of an uncompleted chemical combination, since complete chemical combination is electrically neutral. The phenomena of electrification by splashing shows that this must occur in cases when the two substances are not known to exert any chemical action when tested under ordinary conditions.

We must remember, however, that the layers of substance next the surface are affected by circumstances which are not allowed to exert any influence in ordinary chemical operations. One of the most important differences is the effect which surface tension can exert on chemical actions taking place in the surface layers. The production of a drop of liquid in air involves an increase in potential energy equal to the area of the drop multiplied by the surface tension between the drop and the air; this in the case of water and air would be about 78 ergs per square centimetre of surface of the drop. The production of this energy will be a tax laid upon a thin layer of water whose thickness is comparable with the range of molecular forces: let us take this as 10^{-8}, then 78 ergs per centimetre would if converted into heat be able to raise the temperature of this layer by about 200° C.; a saving then of a fraction in the surface tension might be sufficient to convert the chemical action which effected this saving from one attended by an absorption into one attended by an evolution of heat.

Now Lord Rayleigh[1] has shown that any diminution in the abruptness of transition between two surfaces diminishes the surface tension, and therefore the energy required for the production of the surface. Now, in the case of the drop of water it seems reasonable to suppose that the abruptness of the transition would be mitigated if there was between the water and the air a layer formed of a quasi-compound of the two substances. This layer must not be too thin, for the effect of a transitional layer in diminishing the surface tension diminishes very rapidly with the thickness of the layer when this sinks below a certain value.

The double electric layer must, since it can be separated by mechanical means, have the positive and negative charges separated by a distance much greater than that which separates the positively from the negatively electrified atoms in an ordinary molecule. On the view we have been discussing, the reason the coatings of the electric layer remain at a considerable distance apart is that any further approach of these coatings would, by diminishing the thickness of the transitional layer, increase the potential energy due to surface tension more than it would diminish the potential energy due to the electrostatic attraction of the constituents of the layer.

[1] Rayleigh, Phil. Mag., 33, p. 468, 1892.

With regard to the great effect produced by the addition of a small quantity of some substances to the water, we must remember that the layer whose disruption produces, on this theory, the electrification in the splashing of drops, is due to a partial chemical combination between the liquid and the air. Anything which alters the conditions under which this combination takes place may be expected to alter the electrification produced by the drops. Now the conditions are altered when a foreign substance is added to the water, for the water which is required for any compound of water and air has to be torn from, if not a chemical compound, yet that connection which exists between a salt and a solvent; it may require more work to tear the water away from this connection than could be compensated for by the diminution in the surface tension, and thus the layer might not be formed. In those cases where the addition of the foreign substance reverses the electrification, a layer of a different kind must be formed; this may consist of a combination with the air of the substance added to the water; or possibly a compound into which all three enter.

The magnitude of the effects produced by small quantities of foreign substances is very surprising, but it must be remembered that the chemical actions which produce the layer, such as that between water and the air, are not of a very energetic character, and might therefore be easily affected by apparently trivial circumstances.

The amount of electrification produced by the fall of a drop from a given height depends not merely on the charge of electricity per unit area of the double layer, but also on the ease with which the layers can be separated by the shock produced by the impact of the drop with the plate. This makes the interpretation of the results of experiments on the electrification produced by drops ambiguous, as we cannot tell whether, when we get a big effect, it is due to a highly charged double layer, or to one with a small charge but of which the coatings are exceptionally easily separated.

In connection with this subject it is interesting to find that Kenrick,[1] who measured the potential differences between various liquids and air, found that the substances which when added to water produced a large effect on the potential difference were those which produced a large effect upon the surface tension.

Holmgren[2] has shown that when two wet cloths are rapidly brought together and then pulled

[1] Kenrick, Zeitschrift für physikalische Chemie, 19, p. 625, 1896.

[2] Holmgren, Sur le Developpement de l'électricité au contact de l'air et de l'eau. Société physiographique de Lund, 1894.

quickly apart, electrification is produced, the positive electrification being on the cloth, the negative in the air. He also found that when the area of a surface of water is changing rapidly, — as, for example, when ripples are travelling over it, — there is again electrification, the water being positively, the air negatively, electrified.

The fact that electrification is produced by a sudden diminution in the area of a water surface may be of considerable importance in meteorology; for the coalescence of small drops of water to form a large one would be accompanied by a diminution in the water surface and would give rise to electrification. Thus, as Sir George Stokes has suggested, the large drops of rain which fall in a thunder-storm may be the cause rather than the consequence of the storm.

The production of electrification by the flow of steam through pipes, as in Armstrong's hydro-electric engine, is an example of electrification by the splashing of drops, as Faraday showed that for this electrification to occur it was essential that water-drops should be present in the steam.

Electrification of a Gas by the Aid of Röntgen Rays.

Of all the methods by which we can put a gas into a state in which it can receive a charge of electricity, none is more remarkable than that of the Röntgen rays. These rays, when they pass through a gas, turn it into a conductor and enable it to receive a charge of electricity, and the gas retains its conductivity and its power of being charged for some time after the rays have ceased to pass through it. It may at the outset be worth while to distinguish between the effects of Röntgen rays and those of rays of ultra-violet light. The latter only make the gas a conductor when the light is reflected from a fluorescent substance or from the surface of a metal immersed in the gas; and the gas is only able to discharge a charged body in its neighbourhood which is not illuminated by the ultra-violet rays when the charge on the body is positive. The Röntgen rays, on the other hand, make the gas through which they pass a conductor independently of any reflection of the rays, and the gas when in a conducting state is able to discharge negatively as well as positively charged bodies when it comes into contact with them.

To show this effect, the apparatus used to produce the rays — the Ruhmkorff coil and the exhausted tube — is placed inside a closed iron tank; the iron of the tank stops the rays, so to allow them to get out a hole is cut in the tank, and this hole is covered with a window made of a thin sheet of aluminium which allows the rays to get through. The exhausted tube is put into such a position that the rays emanating from it strike against the window and emerge into the air outside the tank. If now an insulated charged body is placed outside the tank in such a position that when the coil is in action the rays play upon the body, it will be found that even though the insulation is perfect when the coil is not working, yet as soon as the coil is started and the Röntgen rays fall upon the charged body, the charge, whether it be positive or negative, will rapidly disappear.[1]

Next take the insulated body and place it near the tank, but in such a position that the rays do not fall upon it; the insulation will be unimpaired even when the coil is in action; now, with a pair of bellows blow on to the charged body the air from above the aluminium window. This is air through which the Röntgen rays have passed; the charged body will now rapidly lose its charge. The air traversed by the rays has thus retained its conductivity during the time taken by it to pass from the neighbourhood of the aluminium window to the charged body.

To examine in greater detail the properties of

[1] For a qualification of this statement when the initial charges are very small, see p. 54.

gas through which Röntgen rays have passed, the author and Mr. E. Rutherford [1] used the following arrangement. A closed aluminium vessel was placed in front of the window through which the rays passed. A tube through which air could be blown by a pair of bellows led into this vessel; a plug of glass wool was placed in this tube to keep out dust, and a gas meter was placed in series with the tube to measure the rate at which the air passed through. The air left the aluminium vessel through another tube, at the end of which was placed the arrangement for measuring the rate at which electricity leaked through the gas. This was usually a wire charged to a high potential placed in the axis of an earth-connected metal tube through which the stream of gas passed; the wire was connected with one pair of quadrants of an electrometer. The wire was carefully shielded from the direct effect of the rays, and there was no leak unless a current of air was passing through the apparatus; when, however, the current of air was flowing, there was a considerable leak, showing that the air after exposure to the rays retained its conducting properties for the time (about .5 second) it took to pass from the aluminium vessel to the charged electrode.

[1] J. J. Thomson and E. Rutherford, Phil. Mag. [5] 42, p. 392, 1896.

Heating the gas when in this state does not seem to impair its conductivity to any considerable extent; for when we inserted a piece of porcelain tubing between the aluminium vessel and the testing apparatus, we found we could raise the tube to a white heat without affecting the conductivity of the gas, though the gas after coming through the tube was so hot that it could hardly be borne by the hand.

If the gas in its passage from the aluminium vessel to the tester was made to bubble through water, every trace of conductivity seemed to disappear.

The gas also lost its conductivity when forced through a plug of glass wool, though the rate of flow was kept the same as in an experiment without the plug when there was a rapid leak. If the plug was inserted in the system of tubes before the gas reached the vessel where it was exposed to the Röntgen rays, the conductivity was not diminished if the rate of flow was kept constant. These experiments seem to show that the structure in the gas in virtue of which it conducts is such that it is not able to pass through the fine pores in a plug of glass wool. A diaphragm of fine wire gauze or muslin does not seem to diminish the conductivity of the gas passing through it.

The effect of passing a current of electricity

through the gas on its way from the aluminium vessel where it is exposed to the Röntgen rays to the place where its conductivity is tested is very suggestive. This was done by inserting a metal tube in the circuit and fixing along the axis of this tube an insulated wire connected with one terminal of a battery of small storage cells, the other terminal of this battery was connected with the outer tube; in this way a current of electricity was sent through the gas as it passed through the tube. The passage of a current from a few cells was sufficient to diminish greatly the conductivity of the gas passing through the tube and by increasing the number of cells the conductivity of the gas could be entirely destroyed. Thus the peculiar state into which the gas is thrown by the Röntgen rays is destroyed when a current of electricity passes through it.

It is the current which effects this destruction, not the electric field; for when the central wire was enclosed in a glass tube so as to stop the current but maintain the electric field, the gas passed through with its conductivity unimpaired.

The current produces the same effect on the gas as it would produce on a very weak solution of an electrolyte. For imagine such a solution to pass through the tubes instead of the gas, then if enough electricity passed through the solution to decompose all the electrolyte, the solution when it emerged would be a non-conductor, and this is precisely what happens in the case of the gas. The analogy between a dilute solution of an electrolyte and gas exposed to Röntgen rays is complete over a wide range of phenomena.

The fact that the passage of a current of electricity through a gas destroys its conductivity explains a very characteristic property of the conduction of electricity through gases exposed to Röntgen rays; that is, for a given intensity of radiation the current through the gas does not exceed a certain maximum value, whatever the electro-motive force may be,[1]

FIG. 3.

—the current, as it were, gets saturated. The relation between the electro-motive force and the current is shown in the above curve, Fig. 3,

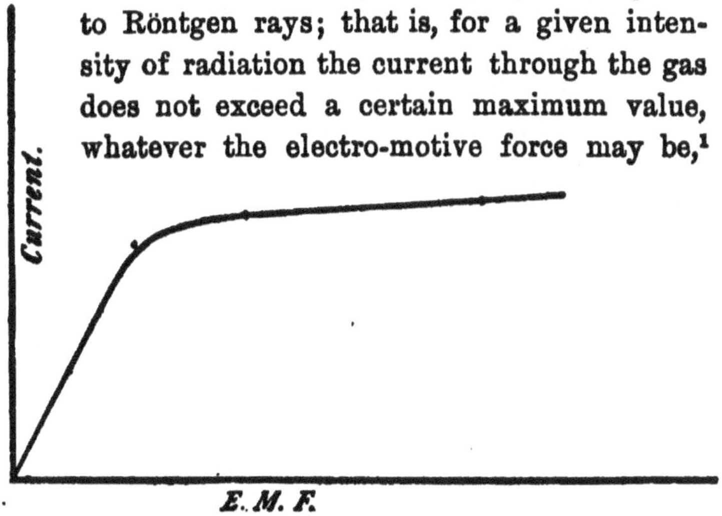

[1] J. J. Thomson, Nature, Apr. 23, 1896.

where the ordinates represent the current, and the abscissæ, the electro-motive forces. For small values of the electro-motive force the curve is straight, showing that the conduction follows Ohm's law; as the electro-motive force increases, the current increases, but at a slower rate than the electro-motive force. The rate of increase of the current gets slower and slower as the electromotive force continually increases, until finally the current reaches a constant value, when an increase in the electro-motive force produces no effect upon the current. It is evident that there must be a limiting value of the current if the passage of the current destroys the conductivity of the gas, and that the maximum current will be the current which destroys the conductivity at the same rate as the Röntgen rays produce it.

If we suppose, as seems proved by some of the preceding results, that the conduction through a gas exposed to the Röntgen rays is electrolytic in character, the rays dissociating the gas and producing a supply of oppositely charged ions, which carry the current, we can easily find an expression connecting the electro-motive force with the current. Let n be the number of ions with charges of one sign in unit volume of the gas, q the rate at which these are produced by the rays; if e is the charge carried by one of the ions, then the passage through the gas of a quantity E of electricity will destroy E/e of these ions. The oppositely charged ions will come into collision and will tend to recombine. The number of recombinations we shall assume to bear a constant ratio to the number of collisions. The number of collisions is proportional to n^2. Let $a\,n^2$ be the number of ions in unit volume which recombine in unit time; let i be the current through unit area of the gas, l the distance between the electrodes. Then we have

$$\frac{dn}{dt} = q - a n^2 - \frac{i}{le}. \qquad (1)$$

So that when the gas is in a steady state,

$$o = q - a n^2 - \frac{i}{le}. \qquad (2)$$

Let us assume that the velocity of the ions is proportional to the potential gradient; let U be the sum of the velocities of the positive and negative velocities when the potential gradient is unity. Then if E is the difference of potential between the plates, the sum of the velocity of the positive and negative ions is $E\,U/l$, and hence

$$i = \frac{nEU}{l}; \text{ or } n = \frac{li}{EU};$$

Substituting this value of n in equation (2), we get

$$o = q - \frac{a\,l^2 i^2}{U^2 E^2} - \frac{i}{le};$$

or
$$q - \frac{i}{le} = \frac{a l^2 i^2}{U^2 E^2}. \quad (3)$$

We see from this equation that when E is infinite, i approaches the value $l\,e\,q$; if we call I the limiting value of i when E is infinite, we have

$$I = q\,l\,e, \quad (4)$$

and from (3)

$$I - i = \frac{a l^2 e}{U^2} \frac{i^2}{E^2}; \quad (5)$$

or writing C for $a l^2 e / U^2$ we get

$$I - i = \frac{C i^2}{E^2} \quad (6)$$

as the relation between the current and the electromotive force. The quantity a is not one that we can measure directly; we shall try to express it in terms of quantities which can be determined experimentally. Let T be the time which elapses after the rays have ceased to pass through the gas before the number of ions falls to one half its initial value, no current passing through the gas in the interval, then we have by equation (1)

$$\frac{dn}{dt} = -a n^2;$$

integrating this equation, we get

$$\frac{1}{n} - \frac{1}{N} = a t, \quad (7)$$

where t is the time which has elapsed since the rays were turned off, and N is the number of ions at the instant when the rays ceased to pass through the gas; we see by equation (1), since no current is passing through the gas in this case, that

$$q = aN^2. \qquad (8)$$

Now when $t = T$, $n = \tfrac{1}{2} N$.
Hence, by (7)

$$\frac{1}{N} = aT;$$

or

$$\left\{\frac{a}{q}\right\}^{\tfrac{1}{2}} = aT.$$

Thus,

$$a = \frac{1}{T^2 q} = \frac{le}{T^2 I}.$$

Making this substitution, equation (5) becomes

$$I(I-i) = \frac{l^4}{T^2 U^2} \frac{i^2}{E^2}. \qquad (9)$$

This equation, when i is small compared with I, its limiting value, becomes

$$\frac{i}{I} = \frac{UT}{l^2} E; \qquad (10)$$

thus the current is proportional to the electro-motive force.

The formula (9) thus coincides with experiment in indicating a limiting value to the intensity of the current, and in making the conductivity obey Ohm's law for small values of the current. The truth of the formula was much more severely tested by the determination by Mr. Rutherford and myself of a long series of determinations of the current corresponding to electro-motive forces varying from 1 or 2 volts to the electro-motive force required to produce saturation. The experiments were made with air, hydrogen, coal gas, chlorine, sulphuretted hydrogen, and mercury vapour; and the intensity of the Röntgen rays as well as the nature of the gas was varied. The results of these experiments are given in the paper by Mr. Rutherford and myself already quoted; the agreement between theory and experiment was very close, never exceeding an amount which might fairly be attributed to errors of experiment. There are some points in the formula which call for special attention. We see from equation (4) that, with a constant intensity of radiation, I, the limiting current, is proportional to l, the distance between the electrodes. Thus, when we approach saturation, the current will increase as the distance between the electrodes increases, and we get what is at first sight the paradoxical result that a thin layer of gas offers a greater resistance to the passage of a current than a thicker one. This is, however, easily explained if we remember that the current destroys the conductivity of the gas, and that as in a thicker layer there are more conducting particles than in a thinner one, the current required to destroy them all will be greater.

The experiments show that the effect of the distance between the electrodes (two parallel plates) on the current is very marked. The following tables show the result of some experiments on this point.

POTENTIAL DIFFERENCE BETWEEN ELECTRODES, 60 VOLTS.

Distance between electrodes in millimetre.	Current (arbitrary scale).
.1	9
.12	15
.25	21
.5	37
1	50
1.5	62
3	91
8	110

The next table contains measurements with a small potential difference of 1.3 volts.

.25	10
.75	32
2	48
3	53
8	53
18	40

We see, as we should expect from the theory, that the effect of distance is not nearly so well marked when the potential difference is small as when it is large.

The measurement of the rate of leak when the current is saturated enables us to form an estimate of the number of ions produced by the Röntgen rays in unit time, as in this case the number of ions produced by the rays is equal to the number destroyed by the current. Let us take the case of an experiment with hydrogen: when the current was saturated, the rate of leak between two plates, each about 10 sq. cm. in area and 1 cm. apart, was in one second about equal to the quantity required to raise a capacity of 30 cm. to the potential of 1 volt. Thus, the quantity of electricity passing between the plates in one second was about 10^{-1} electrostatic units, or $1/3 \times 10^{11}$ electro-magnetic units; and this quantity is sufficient to neutralize all the ions produced in one second by the Röntgen rays. Now 1 electro-magnetic unit of electricity can set free about 10^{-4} grammes of hydrogen, or about 1 cc. of the gas at standard temperature and pressure; hence, $1/3 \times 10^{11}$ units of electricity would be carried by $1/3 \, 10^{11}$ cc. of hydrogen, so that if the ions in the gas carry the same charge as they do in an electrolyte, the volume occupied by the ions produced in one second by the Röntgen rays would only be $1/3 \times 10^{11}$ cc. at standard temperature and pressure. The volume of the gas exposed to the rays was, however, about 10 cc., so that in this experiment the amount of gas ionized was only $1/3 \times 10^{12}$ of the amount of gas exposed to the rays. This result shows that it is not surprising that experiments to see if any alteration in the volume of a gas under constant pressure was produced by the Röntgen rays should have led to negative results. The conductivity of iodine and mercury vapour is much greater than that of hydrogen, so that the number of ions produced by the rays would be greater; but even in the case of the best-conducting vapours the number of ions produced is an exceedingly small fraction of the number of molecules in the gas.

When the current through the gas is small compared with the saturation current, we have by equation (10),

$$\frac{i}{I} = \frac{EUT}{l^2}.$$

Now EU/l is the sum of the velocities of the positively and negatively charged ions in the electric field. Thus, this equation leads to the result that when the current is small, the ratio it has to the maximum current is equal to the ratio of the space passed over by the charged ions in the time T to the space between the electrodes. In an experiment when l was about 1 cm., the rate of leak through air when the potential difference was 1 volt was about 1/30 of the maximum rate of leak; hence, the charged ions must in the time T have moved through about .03 cm. The time T depends upon the intensity of the radiation. A rough estimate gave T as about .1 second in the experiment under discussion; this would make the sum of the velocities of the ions in air, under a potential gradient of 1 volt per cm., equal to .33 cm. / sec.[1] This velocity is very large compared with the velocities of the ions in the electrolysis of liquids; it is, however, small compared with the velocity with which an atom carrying an atomic charge of electricity would move under the potential gradient through a gas at atmospheric pressure. If we calculate by the Kinetic Theory of Gases this velocity, we find that for air it is of the order 50 cm. / sec. The magnitude of this velocity compared with that of the ions seems to show that the ions in a gas exposed to Röntgen rays are the centres of an aggregation of a considerable number of molecules. The result is thus in accordance with the view to which we are led by many other independent considerations, that there is a structure in a gas-conveying electricity whose structure is exceeding coarse compared with a structure consisting of single molecules uniformly distributed.

[1] More accurate experiments have shown that this is about 3.2 cm. per second.

When the intensity of the Röntgen rays is allowed, the alteration in the intensity of the current is not the same at all points on the i and E curve. When the intensity of the rays is changed, the saturation current is increased or diminished in a larger proportion than the current for small

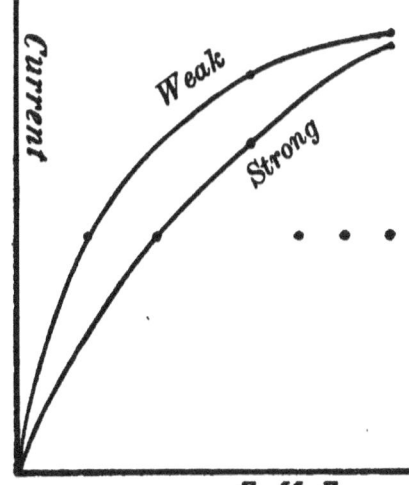

Fig. 4.

electro-motive forces. This is shown by the above diagram, Fig. 4, which represents the i and E curves for conduction through chlorine gas for different intensities of the Röntgen rays; the weak radiation was got by interposing between the bulb and the gas a thick plate of aluminium. In the figure the ordinates for the weak radiation have been increased so as to make the ordinate for the saturation current of the weak radiation the same as that of the strong. When this is done the rest of the "weak" curve is above that of the strong, showing that the diminution in the intensity of radiation has affected the saturation current to a greater extent than the weak current.

These results follow at once from equation (4). I, the saturation current, is given by the equation

$$I = qle,$$

so that it is proportional to the number of ions produced by the rays in unit time. On the other hand, when the current i is small compared with I, we have

$$i = \frac{E}{l} U \sqrt{\frac{q}{a}};$$

thus the small current is only proportional to the square root of q, and therefore does not vary so quickly with q as the saturation current.

If the number of ions produced by the Röntgen rays in a gas were proportional to the number of molecules of the gas in unit volume — that is, to the pressure of the gas, — then q would be proportional to the density of the gas; thus the saturation current would be directly proportional to the pressure, while if the current were a long way from saturation it would be inversely proportional to the square root of the pressure, as U is inversely proportional to the pressure. The results obtained by different observers as to the effect of pressure on the conductivity of a gas exposed to Röntgen rays are not in accordance. M. Perrin, using an electro-motive force sufficiently large to saturate the gas, found that the conductivity was proportional to the pressure; in some experiments made by Mr. McClelland and myself, and in others by M. Humuescu, the conductivity varied as the square root of the pressure. This discrepancy arises in M. Perrin's opinion from an effect produced by the rays at the surface of a metal on which they fall.

Different gases conduct when under the influence of Röntgen rays with very different degrees of facility. Using a constant intensity of radiation, the saturation current for mercury vapour was about twenty times as large as that for air, while that for air was half as large again as the saturation current through hydrogen. Hydrogen has a smaller saturation current than any gas I have tried. Experiments made on the following gases showed that their saturation currents were in the order, Hydrogen, Nitrogen, Air, Oxygen, Carbonic Acid, Sulphuretted Hydrogen, Hydrochloric Acid, Chlorine, Mercury.

The magnitude of the saturation current does not seem to depend altogether upon the density of the gas; thus it is very much greater for H_2S

than for air, although the densities of these gases are nearly equal. Chlorine, Bromine, Iodine, Sulphur compounds and Mercury vapour have large saturation currents, and these elements and their compounds have when in their gaseous state abnormally high specific inductive capacities in comparison with their valencies. The high conductivity of mercury vapour is very remarkable, as this gas is often regarded as monatomic.

Mr. Rutherford[1] has found that the absorption of the Röntgen rays by different gases is proportional to the saturation current through these gases, — that is, if the intensity of the radiation after passing through a stratum of thickness, d, be represented by $e^{-\lambda d}$, then λ is proportional to the saturation current through the gas. If each of the ions carry the same charge, then the saturation current is proportional to the number of molecules of the gas dissociated in unit time, so that another way of expressing this result would be that the dissociation of a molecule into ions causes the same absorption of Röntgen rays, whatever may be the nature of the gas.

The proportion between the conductivities of different gases depends upon the electro-motive force used; thus for small electro-motive forces the conductivity of hydrogen is greater than that of

[1] Rutherford, Phil. Mag.

air, though the saturation current through air is much greater than that through hydrogen. This is shown in the following diagram, which represents the I and E curves for hydrogen and air. The curves intersect, the hydrogen curves being above the air curve for small values of the electromotive force and below it for large values. The saturation current depends merely on the number

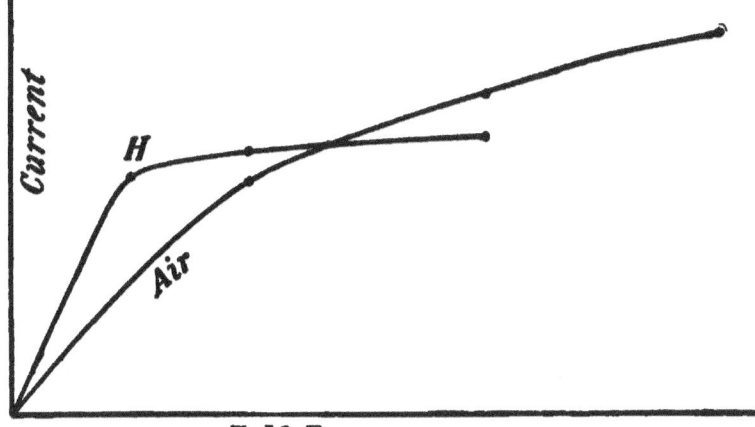

Fig. 5.

of conducting particles produced by the rays while the current in the early part of the curve depends upon the space described by the conducting particles in the time T (see equation 10); and we infer from these curves that more ions are produced by the rays in air than in hydrogen, but that the product of U, the velocity of the ions, and T, a time which is proportional to the time the ions linger after the rays are cut off, is greater for hydrogen than it is for air.

In deducing equation (6), we assumed that the only way in which when a current is not passing through the gas the ions disappeared was by recombining; if the gas is contained in a vessel it may be that some of the ions would give up their charges when they struck against the sides of the vessel. The number so giving up their charges would be proportional to n and not to n^2, and would equal kn where k is a coefficient depending on the shape and size of the vessel. Thus equation (1) would, if this term were included, become

$$\frac{dn}{dt} = q - an^2 - kn - \frac{i}{le};$$

so that the corresponding relation between the current and the electro-motive force would be

$$q - \frac{a}{e^2 U^2}\frac{i^2}{E^2} - \frac{k}{eU}\frac{i}{E} - \frac{i}{le} = o.$$

The effect of the term we are considering would be to diminish the conductivity when the electromotive force is small; it does not affect the limiting value of the current.

Mr. Rutherford[1] has investigated the electrification of a gas exposed to Röntgen rays. The

[1] Rutherford, Phil. Mag., 43, p. 241, 1897.

arrangement used is represented in Fig. 6. The gas was blown rapidly through the tube; after passing out of this tube the gas rushed along the wire placed along the axis of the tube through which the gas was blown into the vessel. The gas as it passed along the wire was exposed to Röntgen rays. After leaving the wire the gas went into

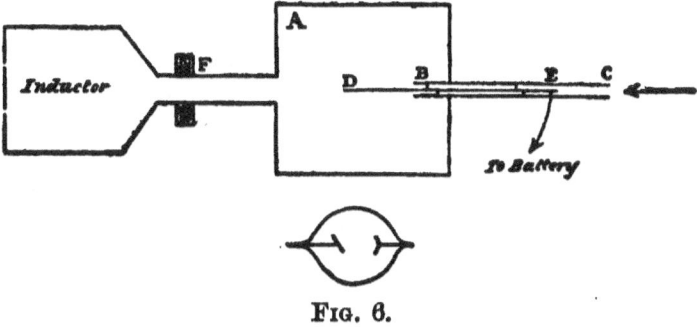

Fig. 6.

the metallic vessel connected with the electrometer. The gas entering this vessel was found to be charged with electricity, and the electrification was of the opposite sign to that on the wire with which the gas had been in contact. This effect admits of a simple explanation on the view that when a gas is exposed to Röntgen rays, positively and negatively charged ions are produced in the gas. Let us suppose that the wire is charged with positive electricity; then it will attract the negatively charged ions in its neighbourhood and repel the positive ones. In the layer of gas next the wire there thus would be an excess of negative over positive ions; and if this layer gets swept past the wire before all the negative ions can strike against the wire and give up their charges, the gas after passing the positively charged wire will have a charge of negative electricity. The greater the velocity of the ions under a given electro-motive intensity, the more likely will the negative ions be to move up to the wire and give up their charges before they are swept past it, and thus contribute nothing to the negative electrification of the gas in the metal vessel.

The fact that the electrification in the gas is of opposite sign to that on the wire seems very strong evidence in favour of the view that there are *charged* ions in the gas exposed to the Röntgen rays, as if the ions carried no charge to begin with, the gas after passing the wire would, if it were charged at all, be charged with electricity of the same sign as that on the wire.

If the charged wire is coated with a layer of paraffin, or placed inside a glass tube so as not to be in actual contact with the stream of gas, the gas is electrified and the sign of the electrification is opposite to that on the wire.

The properties of gases electrified when under the influence of the Röntgen rays differ somewhat from those of gases electrified by other methods.

Thus the electrification in a gas electrified under Röntgen rays cannot pass through glass wool, or bubble through water; it thus resembles the conducting property conferred on the gas by these rays. On the other hand, as we have seen when the electrification in the gas is produced by chemical action, the electrification is able to pass through glass wool, or bubble through conducting liquids.

Another property of the electrification produced in a gas when exposed to Röntgen rays is that it is rapidly taken out of the gas not merely by conductors, but also by insulators in contact with the gas;[1] the electrification cannot survive the passage of a gas through any considerable length of fairly wide bore tubing, whether this be made of metal or glass; while the electrification due to chemical action would not be appreciably affected by the passage through similar tubes. This may be the reason why the electrification produced under the rays cannot pass through plugs of glass wool, or bubble through water or sulphuric acid.

Mr. Rutherford also found that the negative ions gave up their charges more readily to metals than the positive ions. The difference was more marked with some metals than with others; thus the difference was greater for zinc than for copper. In fact, the more electro-positive the metal, the more

[1] Rutherford, Phil. Mag., 43, p. 241, 1897.

did the rate at which the negative charge was given up exceed that for the positive. No difference of this kind was observed in the case of insulators. Mr. Erskine-Murray [1] found that a gas exposed to the Röntgen rays produced the same effect on the potential difference between two metals immersed in the gas as is produced when the two metals are connected by an electrolyte; that is, the two metals are brought to the same potential, in the sense that if, being connected by the conducting gas, the rays are cut off so that the metallic plates are insulated from each other, no charge of potential is produced by separating the plates. If the plates had been connected by a metallic conductor and then disconnected, the difference of potential between the plates would increase as the distance between them was increased. Lord Kelvin [2] has shown that the flame gases arising from the combustion of a spirit lamp produced a similar effect on the potential difference between two metals to that produced by the Röntgen rays.

Professor Minchin has shown that metal plates previously uncharged get, when exposed to Röntgen rays, charges of electricity, positive in some cases, negative in others. If a metal which when originally uncharged gets negatively electrified when

[1] Erskine-Murray, Proc. Roy. Soc., 49, p. 333, 1896.
[2] Kelvin, Reprint of Papers on Electrostatics and Magnetism.

exposed to the rays is charged with negative electricity, it will not lose the whole of its charge; while, on the other hand, if it is initially charged positively, it will not only lose its positive charge, but will acquire a small negative one. The potentials to which these residual charges raise the metals are not large, being generally much less than a volt, except in the case of sodium or (what is much more convenient to work with) sodium amalgam, when it may amount to several volts. The charge assumed by a metal plate will probably depend on the nature of the metals with which it is connected by gas traversed by Röntgen rays. If these are electro-negative to the metal under consideration, it will get a negative charge; if, on the other hand, they are electro-positive to it, it will get a positive charge. We may look upon the system of metals and Röntgenized air as analogous to metals immersed in an electrolyte and forming a galvanic battery; in this case, if the electro-positive metal is connected to one quadrant of an electrometer, it will show a negative charge if the other metal and the other quadrant of the electrometer are connected with the earth.

Perrin has lately discovered that the Röntgen rays, in addition to the effect they produce on the gas through which they pass, exert a special effect on the gas in the neighbourhood of a metallic surface on which they impinge. The layer of gas adjacent to the metal acquires an abnormally great conductivity. The magnitude of the effect varies very much with the nature of the metal and of the gas; it is very small for aluminium, but considerable for gold, zinc, lead, and tin.

Uranium Radiation.

The salts of uranium as well as the metal itself were found by Becquerel to emit something which produced similar effects to those produced by the Röntgen rays. The radiation from uranium, like the Röntgen radiation, can affect a photographic plate after it has passed through films of thin metal which are opaque to ordinary light; it also, like the Röntgen radiation, renders a gas through which it passes a conductor of electricity. The gas through which the uranium radiation has passed has properties quite analogous to those of gas through which the Röntgen rays have passed. Thus the gas retains its conducting properties for some time after the rays have ceased to pass through it; the laws of conduction — that is, the connection between the electro-motive force and the current — are the same in the two cases. The current gets "saturated" and does not increase beyond a certain point, however much the electric intensity is increased; the saturation current is greater for a thick layer of the gas than for a thin one. The conducting property is destroyed by passing the gas through glass wool, or by sending a current of electricity through it. Again, the rate of leak through different gases varies; it is greatest for those gases which give a large rate of leak under Röntgen rays. Mr. Rutherford has lately found that the velocity of the ions when a gas is traversed by uranium rays is the same, if the electric field is the same, as when it is traversed by Röntgen rays. The resemblance between the conduction through gases under Becquerel radiation and under Röntgen radiation is so complete as to make it almost certain that the conduction is produced by the same mechanism. We have seen that the hypothesis that the gas is ionized when exposed to Röntgen rays explains the laws of conduction in that case; we therefore conclude that a gas is ionized by the uranium radiation.

Becquerel has shown that the uranium compounds emit this radiation apparently with undiminished energy, even though they have been kept in the dark for months before they are tested. He found, moreover, that solutions of uranium salts emit this radiation.

Becquerel found that this radiation could be reflected, refracted, and polarized, so that it is evidently a form of light. He found that the refractive index of glass for this light was not very different from the value for ordinary light. If this result should be confirmed, it would seem to show that the wave-length of the uranium light was not an excessively small fraction of that of ordinary light, as on all theories of dispersion the refractive index for light of an infinitely small wave-length is unity.

PHOTO-ELECTRIC EFFECTS

PHOTO-ELECTRIC EFFECTS

THE discovery by Hertz,[1] in 1887, that the incidence of ultra-violet light on a spark gap facilitates the passage of a spark, led to a series of investigations by Hallwachs,[2] Hoor,[3] Righi,[4] and Stoletow[5] on the effect of ultra-violet light on electrified bodies. It was found that a freshly cleaned surface of zinc, if charged with negative electricity, rapidly loses its charge, however small this may be, when ultra-violet light falls upon it; whereas a similar surface charged with positive electricity suffers no loss of charge, and further that the surface if uncharged acquires a positive charge when exposed to the light. The ultra-violet light may be obtained from an arc-lamp, the effect of which is increased if one of the terminals is of

[1] Hertz, Wied. Ann., 31, p. 983, 1887.
[2] Hallwachs, Wied. Ann., 33, p. 308, 1888.
[3] Hoor, Repertorium der Physik, 25, p. 105, 1889.
[4] Righi, C. R., 107, p. 560, 1888.
[5] Stoletow, C. R., 106, pp. 1149, 1593; 107, p. 91; 108, p. 1241.
See also Physikalische Revue, bd. 1, 1892.

zinc or aluminium, the light from these substances being very rich in ultra-violet rays; it may also be got from burning magnesium, or by sparking with an induction coil between zinc or cadmium terminals. Ordinary sunlight is not rich in ultra-violet light, and does not produce anything like so great an effect as the arc-light. Elster and Geitel,[1] who have investigated the effects of light on electrified bodies with great success, found that the more electro-positive metals lose negative charges when exposed to ordinary light and do not require the presence of the ultra-violet rays. They found that amalgams of sodium and potassium enclosed in a glass vessel lose a negative charge when exposed to daylight, though the glass stops the small amount of ultra-violet light left in sunlight after its passage through the atmosphere. If sodium or potassium, by themselves, instead of their amalgams, be used, or (what is often more convenient) the curious liquid obtained by mixing sodium and potassium in the proportion of the combining weights of these metals, they found that the photo-electric effects are produced by the light from an ordinary petroleum lamp. While if the still more electro-positive metal, rubidium, is used, the photo-

[1] Elster and Geitel, Wied. Ann., 38, pp. 40, 497, 1889; 41, p. 161, 1890; 42, p. 564, 1891; 43, p. 225, 1892; 52, p. 433, 1894; 55, p. 684, 1895.

electric effects due to the light from a glass rod just heated to redness can be distinctly perceived. They found, however, that the eye was more sensitive to radiation than the rubidium cell, and no photo-electric effects could be detected before the radiation from the glass rod was visible.

Elster and Geitel arrange the metals in the following order for their photo-electric effects:

>Rubidium.
>Potassium.
>Alloy of Sodium and Potassium.
>Sodium.
>Lithium.
>Magnesium.
>Thallium.
>Zinc.

With copper, platinum, lead, iron, cadmium, carbon, and mercury, the effects with ordinary light are too small to be appreciable. This order is the same as that in Volta's electro-chemical series.

Elster and Geitel found that the ratio of the photo-electric effects of two metals exposed to approximately monochromatic light depended upon the wave-length of the light, different metals exhibiting a maximum sensitiveness in different parts of the spectrum. Thus, in a solar spectrum obtained by a glass prism, the blue is the part which produces the greatest effect on potassium. The following table for the alkaline metals, given by Elster and Geitel,[1] shows how the photo-electric effect for a particular metal depends upon the character of the incident light. The numbers in the table are the rates of emission of negative electricity under similar circumstances. The rate of emission for each of the cells under the white light from a petroleum lamp is taken as unity.

	Blue.	Yellow.	Orange.	Red.
Rb	.16	.64	.33	.039
K	.57	.07	.04	.002
Na	.37	.36	.14	.009

This table indicates that the absorption of light by the metal has a great influence on the photo-electric effect, for while potassium is more sensitive to blue light than sodium, the strong absorption of yellow light by sodium makes it more than five times more sensitive to this light than potassium. Stoletow very early called attention to the necessity of strong absorption for photo-electric effects. He showed that water, which does not absorb the visible or ultra-violet rays much, does not lose a charge of negative electricity when illuminated, while strongly coloured solutions and especially solutions of fluorescent substances, such as methyl green and methyl

[1] Elster and Geitel, Wied. Ann., 52, p. 438, 1894.

violet, do so to a very considerable extent; he found indeed that a solution of methyl green was very much more sensitive than zinc. Phosphorescent substances, such as Balmain's luminous paint (sulphide of calcium) show this photo-electric effect, so also, as Elster and Geitel [1] have shown, do various specimens of fluor-spar, the magnitude of the effect depending to a very great extent on the colour of the spar. As phosphorescence and fluorescence is probably accompanied by a very intense absorption by the surface layers of the substance, the evidence seems very strong that in order to get the photo-electric effect there must be strong absorption of some kind of light, whether this be ultra-violet light or light of longer wave-length. Hallwachs [2] has shown that, in the case of liquids, there is always strong absorption whenever the liquid shows photo-electric effects; we may have, however, strong absorption without photo-electric effects. When there is strong absorption of light of a particular frequency, say p, we should be led by all theories of dispersion to believe that abnormally great effects on the surface would be produced by all light vibrations having a frequency between p and $p + m$, where m is a positive finite quantity depending on the nature of the absorbing substance.

[1] Elster and Geitel, Wied. Ann., 44, p. 722, 1891.
[2] Hallwachs, Wied. Ann., 37, p. 666, 1889.

Let us take Helmholtz'[1] theory of dispersion as an example. The intense absorption of light is due to the frequency of the absorbed light being equal to one of the free periods of the substance. Helmholtz gives the relation between the frequency of vibration and the refractive index of a substance which has one free period of vibration. This relation is represented by the curve, Fig. 7,

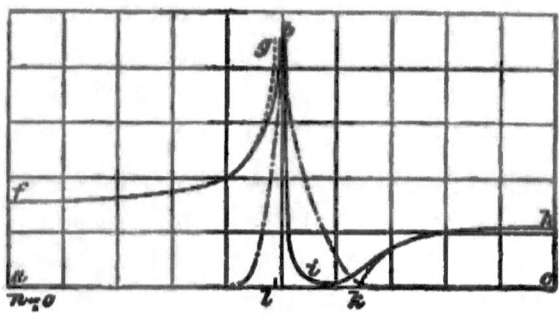

Fig. 7.

in which the ordinates represent the refractive index, and the abscissæ the frequency of the vibration; the refracting substance is supposed to have a free period whose frequency is represented by al. It will be noticed that there is a gap in the curve just after passing this frequency; thus light whose frequency is between al and ak cannot pass through the medium. This light will be totally reflected from the surface, and this total reflection

[1] Helmholtz, Collected Works, vol. iii., p. 505.

will be accompanied by an intense agitation of the molecules in a very thin layer close to the surface. Though the preceding curve only applies to the extreme case when the refracting substance has only one free period, yet when there are many free periods the effects, though more complicated, will probably be of the same general character. It is

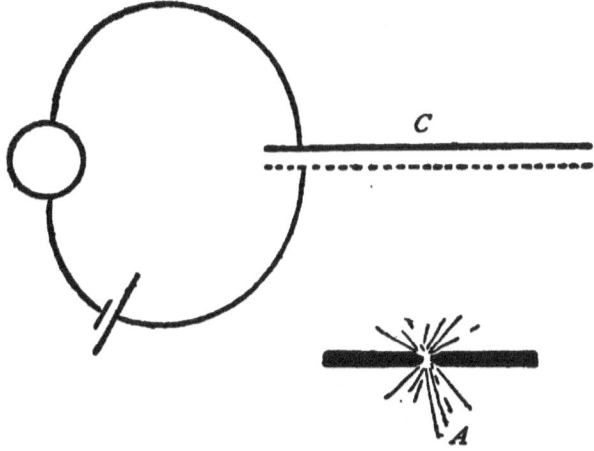

FIG. 8.

this intense agitation of the surface layer — that is, the possession of the molecules in this layer of an abnormal amount of energy — which seems necessary for the production of photo-electric effects.

The laws which regulate the flow of negative electricity from the illuminated surface are very interesting. They have been investigated by Stoletow,[1] Righi,[2] and Elster and Geitel.[3] The apparatus used by Stoletow is shown in Fig. 8. The light from an arc-lamp, A, passed through a hole in a metal screen; it then fell on the metal plates, C, connected with a battery and galvanometer in the way shown in the figure. The plate nearest the light was perforated, and the light passed through the perforations on to the sensitive plate connected with the negative pole of the battery. The current passing between the plates was measured by the galvanometer. By means of this apparatus Stoletow investigated the relation between the current and the potential difference between the plates, varying in his experiments the distance between the plates; his results are represented by the curves in Fig. 9. The curves show that except for small electro-motive forces the conduction does not obey Ohm's law; the current does not increase as rapidly as the electro-motive force. For large values of the electro-motive force the curves give indications of becoming straight lines slightly inclined to the axis along which the electro-motive force is measured. In conduction through a gas illuminated by Röntgen rays, these lines become parallel to the

[1] Stoletow, Journal de Physique, (2) 9, p. 468–473, 1890.
[2] Righi, Mem. della R. Acc. de Bologna, (4) 10, pp. 85–114, 1890.
[3] Elster and Geitel, Wied. Ann., 52, p. 438, 1894.

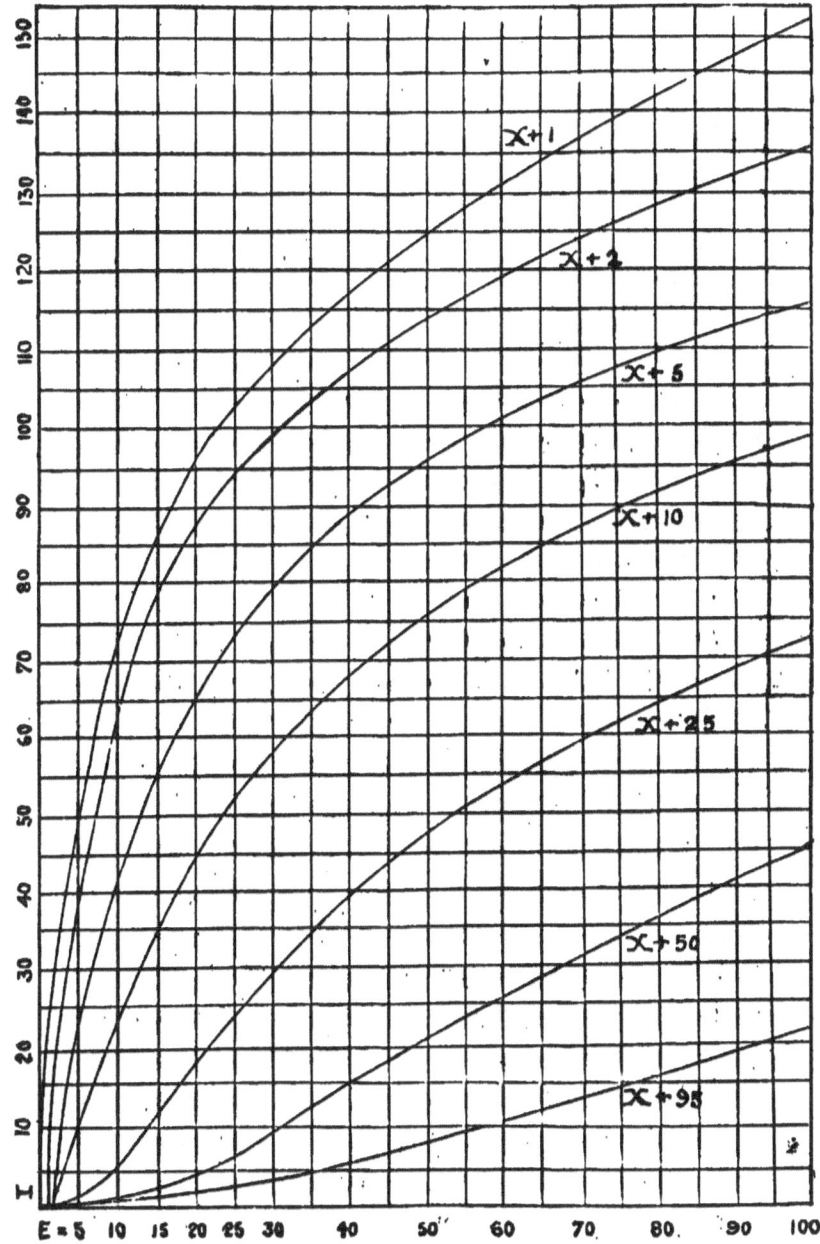

axis. Stoletow and Righi have investigated the effect of the pressure of the gas on the rate of leak. They found that the current increased as the pressure diminished until a certain pressure was reached, when any further diminution in the pressure caused a diminution in the current. The change in the current was small compared with the change in the pressure. The following are some measurements by Stoletow with an electromotive force equal to that of 65 Clark cells. The distance between the plates was 3.71 mm.; p stands for the pressure, and c for the corresponding current.

$p =$	754	152	21	8.8	3.3	2.48	1.01
$c =$	8.46	13.6	26.4	32.2	48.9	74.7	106.8
$p =$	0.64	0.52	0.275	0.105	.0147	.0047	.0031
$c =$	108.2	102.4	82.6	65.8	53.8	50.7	49.5

If p is the pressure at which the current is a maximum, then Stoletow states that $p\,l/E$ is constant, where E is the electro-motive force and l the distance between the plates; according to Righi p is the pressure at which sparks pass most easily between the plates.

The nature of the gas between the plates has a considerable effect upon the rate of leak. Elster and Geitel[1] investigated the rate of leak from an illuminated surface through air, carbonic acid

[1] Elster and Geitel, Wied. Ann., 41, p. 161, 1890.

gas, oxygen, and hydrogen; they found that the rate of leak through carbonic acid was much faster than for any of the other gases.

Elster and Geitel [1] found that the plane of polarization of the incident light had considerable influence upon the amount of the photo-electric effect. The effect produced by light polarized in a plane at right angles to the plane of incidence is for oblique incidence much greater than that produced by light polarized in the plane of incidence. This effect is much more marked in the case of the liquid alloy of potassium and sodium surrounded by gas at a low pressure and exposed to visible light than it is for a zinc plate surrounded by air at atmospheric pressure and exposed to ultra-violet light.

In light polarized in a plane at right angles to that of incidence, the periodic electric intensity, which, according to the electro-magnetic theory of light, exists in the incident light wave, has a component at right angles to the reflecting surface, and produces a periodic effect on the electrification of the surface. When the light is polarized in the plane of incidence, the electric intensity is parallel to the reflecting surface and the electrification over this surface has no tendency to a periodic variation.

[1] Elster and Geitel, Wied. Ann., 55, p. 684, 1895.

The existence of a periodic electric intensity at right angles to the reflecting surface seems to facilitate the escape of negative electricity from the surface. Another circumstance which may be connected with the effect produced by the position of the plane of polarization is the fact discovered by Quincke,[1] that light polarized at right angles to the plane of incidence penetrates more deeply into metals than light polarized in that plane.

Elster and Geitel[2] also found that the rate of escape of negative electrification from a charged surface is diminished by a magnetic field when the lines of magnetic force are parallel to the reflecting surface. This effect is small at ordinary pressures, but increases rapidly as the pressure diminishes. In one of the cases investigated by Elster and Geitel, when the gas was oxygen at a pressure of .0002 mm. of mercury, the rate of leak in a magnetic field (whose strength is not given) was only half the rate when there was no magnetic force.

Stoletow,[3] Righi,[4] and Arrhenius[5] have shown that when two different metals, M and M^1, are used

[1] Quincke, Pogg. Ann., 129, p. 177, 1866.
[2] Elster and Geitel, Wied. Ann., 41, p. 166, 1890.
[3] Stoletow, Physikalische Revue, 1, p. 765, 1892.
[4] Righi, Journal de Physique, (2) 7, p. 153, 1888.
[5] Arrhenius, Wied. Ann., 33, p. 638, 1888.

for the perforated and continuous plate in an arrangement like that in Fig. 8, the total electromotive force acting round the circuit is not E, the electro-motive force of the battery, but $E \pm M/M^1$, where M/M^1 is, approximately at any rate, the contact difference of potential between the metals when these are not exposed to ultra-violet light. Thus the air traversed by ultra-violet light behaves like an electrolyte. Stoletow has verified by direct experiment that two different metals when immersed in the illuminated air are at the same potential, just as Lord Kelvin has shown they are when immersed in an electrolyte, and as they are when immersed in air near a flame or in air traversed by Röntgen rays.

When the plates are of the same material, there is no polarization produced by the passage of a current through the illuminated gas.

Stoletow constructed a battery of two parallel plates made of different metals; one of these was perforated, and ultra-violet light passed through the openings and fell upon the other plate. For this battery to work, it is necessary that the perforated plate should be made of the more electro-positive metal, as in this case the current goes round in such a direction that the negative electricity passes from the plate on which the ultra-violet light shines to the perforated plate. If the plates were reversed then for the current to pass, the illuminated plate would have to discharge positive electricity, and this, as we have seen, it is unable to do.

Stoletow[1] made a series of experiments to see if the discharge from a negatively electrified plate lasted for an appreciable time after the ultra-violet light was cut off. He was not able to obtain any evidence that the cessation of the discharge was not absolutely contemporaneous with the withdrawal of the light, and was able to prove that there was no leakage after the rays had been cut off for 1/1000 of a second.

The investigations by Lenard and Wolf[2] on the behaviour of a steam jet placed near a surface illuminated by ultra-violet light are of great interest in connection with these photo-electric effects. These observers found that when ultra-violet light falls on a negatively electrified platinum surface, a steam jet in the neighbourhood of the surface shows by its change of colour that the steam in it has been condensed. The metals tried by Lenard and Wolf were zinc, mercury, platinum, brass, copper, tin, lead, gold, and silver, and in all these cases condensation of the steam occurred when the metallic surface was negatively electri-

[1] Stoletow, Physikalische Revue, 1, p. 747, 1892.
[2] Lenard and Wolf, Wied. Ann., 37, p. 443, 1889.

fied. They found too that though no condensation is produced when the light falls on a water surface, yet it occurs when the water surface is replaced by one of the fluorescent solutions, such as rosaniline or methyl violet, which give photo-electric effects. They found also that condensation occurs, though to a much smaller extent, if the surfaces are not electrified; but no condensation can be detected when the surfaces are charged with positive electricity. It will be noticed that some of the metals which are effective in producing condensation are not very sensitive to the photo-electric effect, and surfaces of quartz or gypsum also produce condensation; glass and mica, on the other hand, give no appreciable effect.

Lenard and Wolf attributed this condensation of the steam in the jet to dust emitted from the illuminated surface, the dust, in accordance with Aitken's experiments,[1] producing condensation by forming nuclei around which the water-drops collect.

The indications of a steam jet are, however, very ambiguous, as condensation is promoted not only by dust, but also by chemical action or electrification in its neighbourhood. Thus Lenard and Wolf's experiments do not determine whether metallic dust or the gas is the carrier of the electrification.

[1] Aitken, Trans. Roy. Soc., Edinburgh, 30, p. 337, 1887.

Lenard and Wolf observed a slight roughening of the metal at the places exposed to the ultra-violet light.

As the direct evidence as to the nature of the carriers of the electrification in these photo-electric effects is inconclusive, we must have recourse to the indirect evidence afforded by the laws obeyed by the convection currents which start from the illuminated surface; those seem to point to the gas playing a very considerable part in the discharge. The fact that the rate of discharge depends upon the nature of the gas is not conclusive, as, if the discharge were carried by the metallic dust, we might expect that the rate of diffusion of the dust through the surrounding gas would vary with the nature of the gas. The way in which the rate of leak varies with the pressure seems, however, inconsistent with the idea that the metallic dust carries the greater part of the charge. As we diminish the pressure the rate of leak does not continually increase; it increases indeed until the pressure sinks to a certain critical value, but when the pressure falls below this value, any further decrease in the pressure produces a decrease in the rate of leak. This seems to point to the gas as the carrier of the greater part of the convective current. It is, however, to be remembered that the gas is not put into the state in which it can act as a carrier of electricity merely by the passage through it of ultra-violet light; this state of the gas is manufactured only when the light strikes against the surface of the electro-positive metal or the solution of a phosphorescent substance. Its production depends on the material of which the reflecting surface is made, and even, when the light is polarized, on the orientation of this surface. It therefore differs essentially from the conductivity produced in a gas by the passage through it of Röntgen rays, which is not dependent upon the existence of suitable surfaces for these rays to strike against.

The sign of these photo-electric effects is opposite to that produced when the surface is made so hot as to be luminous; for in air, as Elster and Geitel have shown, the hot metal becomes negatively, the surrounding air positively, electrified. The sign of the electrification in the air produced by the action of light is the same as that produced in air by the splashing of drops of mercury or water; and as the electrification, in that case, was explained by the existence of a double layer of electrification, the positive side of which was on the metal, the negative on the gas, the negative layer getting partially torn away by the shock produced by the splash, so it would seem that the photo-electric effects might be produced by the tearing away of the outer layer of the double sheet by the action of the incident light. There seem reasons for thinking that the incident light might produce this result; the photo-electric effects are accompanied by an intense absorption of light at the surface layer of the reflecting body, so that a considerable amount of energy will flow into the molecules of this layer, which forms one of the sides of the electrical double layer at the surface. These particles, having a large amount of kinetic energy, may by communication so increase the kinetic energy of the gaseous particles which form the opposite side of the double layer, that it gets sufficiently large to enable some of them to escape from the neighbourhood of the reflecting surface and diffuse into the surrounding gas; as the particles of gas in the outer layer are negatively charged, a negative charge will diffuse into the gas and a positive one be left on the metal. If the metal itself is charged negatively, the repulsion exerted by this charge on the negatively electrified particles of gas will make these move more quickly and so accelerate the rate of leak; a positive charge, on the other hand, would retard it. The effect of pressure on the rate of leak is consistent with this view, for reduction of pressure would, on the one hand, facilitate the leak by increasing the rate of diffusion of electrified particles through it; on the other hand, it would retard the rate of leak if it affected the number of particles available for forming the outer layer; it would seem likely that the pressure would have to be very low before much effect was produced on the number of charged particles in the outer layer, so that it would not be until the pressure is very low that a decrease in the pressure would cause a decrease in the rate of leak. The action of a magnetic field on the rate of leak is also consistent with this view, as a magnetic force acting parallel to the reflecting surface would act upon the moving charged particles with a force acting at right angles both to the direction of motion of the particles and to the magnetic force. This force will leave the velocity unaltered, but will constrain the particles to follow a curved path; this will diminish the component of the velocity parallel to the electric intensity and so diminish the rate of leak.

This view also affords an explanation of the curious connection between the rate of leak and the electro-motive force illustrated by Fig. 9.

This view seems also to explain the existence of a pressure at which the current is a maximum. For we may suppose that the effect of a diminution in pressure is twofold: the first an effect on the velocity of the electrified particles through the gas, the velocity varying inversely as the pressure;

the second an effect upon the number of active particles produced by the rays, the number diminishing when the pressure is reduced, but varying more slowly than the pressure. Starting with a higher pressure, the effect of diminution of the pressure will be to increase the velocity with which the particles move, in a greater ratio than it diminishes the number of particles; thus we shall get an increase in the current. This will go on until the current is so great that to carry it requires all the active particles produced by the ultra-violet light; as soon as this is the case, the current depends only upon the number of particles produced, and not upon their velocity. Any further diminution of the pressure will therefore produce a diminution in the current, since it diminishes the number of active particles. There will thus be a certain pressure when the current is a maximum.

Stoletow found that at this pressure an increase in the potential difference produced an increase in the current. This shows, if the preceding explanation is the correct one, that the greater the density of the negative electricity on the plate on which the ultra-violet light fell, the greater the number of active particles produced; this would happen if the effect of the ultra-violet light was to produce a definite difference of potential between the gas and the metal.

A very interesting result obtained by Stoletow was that, for very low pressures, the maximum current was independent of the pressure; this would seem to indicate that the current in these cases was carried either by the mercury vapour from the pump or by particles from the metal surface.

Electrification of Gases by Glowing Metals.

Elster and Geitel [1] found that when a platinum wire is heated to luminosity in air, the air near the wire acquires a charge of positive electricity, the wire itself a charge of negative electricity; if, however, the wire is heated in hydrogen, the electrification in the hydrogen is negative, that on the wire positive.

Hydrogen was the only gas amongst those they tried (oxygen, carbonic acid gas, water vapour, and the vapours of sulphur and phosphorus and mercury) which acquired a negative charge; the others all got charged positively, with the exception of mercury vapour, which did not get charged at all. The same effects were observed when palladium and iron wires were used instead of platinum. When carbon filaments were heated, the electrifica-

[1] Elster and Geitel, Wied. Ann., 16, p. 193, 1882; 19, p. 588, 1883; 22, p. 123, 1884; 26, p. 1, 1885; 31, p. 109, 1887; 37, p. 315, 1889. (This paper contains a summary of their results.)

tion in the gas was always negative; these filaments, however, give off so much gas that the conditions of the experiments become indefinite.

In these experiments precautions were taken to get rid of all dust from the gas; but though the gas may be freed from dust at the beginning of the experiment, yet inasmuch as glowing metals give off either metallic dust or vapour, the gas is liable to get charged with metallic dust as the experiment goes on. According to Nahrwold,[1] little if any metallic dust is given off when platinum is raised to incandescence in hydrogen.

If air is blown past an incandescent platinum wire, it comes off positively electrified; thus heating the wire to redness, and blowing air in its normal electric condition past the wire, produces electric effects of the same sign as when the wire is cold, and air through which Röntgen rays have recently passed is blown past it. If another (cold) platinum wire is placed in the neighbourhood of the one raised to luminosity, its potential will be raised in consequence of the positive electrification in the air around it. The potential attained by this wire seemed almost independent of the pressure of the gas. It depends, however, to a great extent on the temperature of the incandescent wire, the potential attaining a maximum at a

[1] Nahrwold, Wied. Ann., 35, p. 107, 1888.

bright yellow heat, after which any increase in temperature causes a diminution in the potential. In hydrogen, where the electrification is negative, an increase in the temperature seems always to be accompanied by an increase in the negative electrification of the gas. At very high temperatures the positive electrification in air gets exceedingly small; indeed, some observers have thought that at very high temperature they detected a tendency for the electrification in the air to become negative instead of positive; and thus to have the same sign as air in the neighbourhood of a clean metallic surface reflecting ultra-violet light. When very thin wires are used and the pressure is very low, then a conductor in the neighbourhood gradually acquires, after the glow has lasted for a long time, a negative charge. The wires become brittle, and their resistance is changed.

Branly[1] investigated this effect of glowing bodies by a slightly different method: he suspended near the glowing body an insulated, charged conductor, and observed whether this was discharged or not. Where the glowing body was a platinum spiral, he found that when this was at a dull red heat, a neighbouring conductor lost a charge of negative electricity, but not of positive. This is what we should expect from Elster and

[1] Branly, C. R., 114, p. 1531, 1892.

Geitel's result, that there is positive electrification in the air around the glowing body. When the platinum spiral was made to glow very brightly, Branly found that the conductor lost its charge, whatever the sign of it might be. He found that the effects at a dull red heat depended upon the nature of the glowing body. Where this was a lamp-shade covered with bismuth oxide or lead oxide, he found that this discharged a positively electrified body in its neighbourhood, but not a negatively electrified one. This is a reversal of the effect observed with clean metallic surfaces.

Mr. Stanton [1] found that a hot surface of clean copper discharges a negatively electrified conductor in its neighbourhood, but ceases to do so when the copper surface gets coated with a layer of oxide; thus the discharge goes on as long as oxidation is taking place, but ceases as soon as this stops. When, however, the hot oxidized copper and the positively charged conductor are placed in a vessel containing hydrogen, so that the oxide gets reduced, the conductor is discharged as long as the reduction is going on, but ceases as soon as the reduction is completed, so that hot, clean copper in hydrogen is unable to discharge a conductor placed in its neighbourhood if this is positively charged; on the other hand, the con-

[1] Stanton, Proc. Roy. Soc., vol. xlvii., p. 559, 1889.

ductor is discharged if it originally has a negative charge. Thus, in hydrogen, hot copper can retain a charge of negative, but loses one of positive, electricity.

Another aspect of the electrification produced by glowing bodies is what is known as "unipolar conduction," — that is, when a hot body loses electricity of one sign more easily than it does that of the opposite sign. Thus a hot platinum wire in air, since it, if unelectrified to begin with, acquires a negative charge, will evidently lose a positive charge more quickly than a negative one. This difference between the rate of escape of the two electricities from hot bodies was known before the production of electrical separation by glowing metals was directly demonstrated. Guthrie,[1] who was the first to call attention to phenomena of this kind, observed that an iron sphere in air cannot, when white hot, retain a charge either of positive or of negative electricity, and that as it cools it acquires the power of retaining a negative charge before it can retain a positive one. If the sphere is connected with the earth and held near a charged body, then, when the sphere is white hot, the body soon loses its charge, whether this be positive or negative; when the sphere is somewhat colder, the body is discharged when negatively, but not when positively, electrified.

[1] Guthrie, Phil. Mag., [4] 46, p. 257, 1873.

Elster and Geitel[1] made the very interesting observation that at low pressures the positive electrification in the case of air is increased by a magnetic field, while the negative electrification in hydrogen is diminished; the latter result is the most marked of the two.

Electrification in the Neighbourhood of an Arc Discharge.

The author has found that electrical effects, very similar to those produced by incandescent solids, occur near the arc. The following arrangement was used for these experiments. An arc discharge between the platinum terminals A, B (Fig. 10), was produced by a large transformer, which transformed up in the ratio of 400 to 1. A current of about 40 amperes, making 80 alternations per second, was sent through the primary. When the gases were at atmospheric pressure the method used was as follows. A current of the gas under examination entered the vessel through a glass tube, C, and blew the gas in the neighbourhood of the arc against the platinum electrode, E, which was connected with one quadrant of an electrometer, the other quadrant of which was connected with the earth. To screen E from external electric influences, it was enclosed in a platinum tube,

[1] Elster and Geitel, Wied. Ann., 38, p. 27, 1889.

ELECTRIFICATION OF GASES 87

which was closed in by fine wire gauze, which, though it acted as an electric screen, yet allowed gases in the the neighbourhood of the arc to pass through it. This tube was connected with the earth. The experiments were of the following kind. The quadrants of the electrometer were

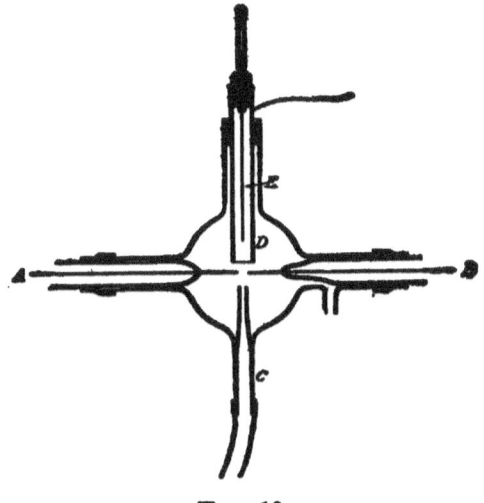

FIG. 10.

charged up by a battery; the connection with the battery was then broken, and the rate of leak observed. When the arc was not passing, the insulation was practically perfect. As soon, however, as the arc was started, and as long as it continued, the insulation of the gas gave way. There are, however, some remarkable exceptions to this. When the gas through which the arc passed was oxygen, the electrode E, if negatively charged to begin with, lost its charge very rapidly; it did not, however, remain uncharged, as it acquired a positive charge, which increased until E acquired a potential V. V depends greatly upon the size of the arc and the proximity to it of the electrode E. In many of the experiments it was as large as 10 or 12 volts.

When E was initially charged positively to a high potential, the electricity leaked from it until the potential fell to V; when this potential was reached, the leak stopped and the gas seemed to insulate as well as when no discharge passed through it. If the potential to which E was raised initially was less than V (in particular, if it was without charge to begin with), the positive charge increased until the potential of E was equal to V when it remained constant. Thus we see (1) that an electrode immersed in the oxygen near the arc can retain a small positive charge perfectly, while it very rapidly loses a negative one; (2) that an uncharged electrode placed near the arc acquires a positive charge. Thus the air in the neighbourhood of the arc shows the same properties as the air in the neighbourhood of an incandescent platinum wire.

The Arc in Hydrogen.

When similar experiments were tried in hydrogen, the results were quite different. When the *arc* passed through hydrogen, the electrode E always leaked when it was positively electrified; it did not remain uncharged, but acquired a negative charge, — the effects being thus of opposite sign to those in oxygen. Here again the electrification in the hydrogen in the neighbourhood of the arc is the same as those in the neighbourhood of an incandescent platinum wire. It is much more difficult to get a good arc in hydrogen than it is in oxygen, and the effects in hydrogen are less marked and are more troublesome to get than those in oxygen.

Experiments were also made with the arc passing through gases at pressures less than the atmospheric pressure; in these experiments the blast had to be dispensed with, the terminals of the arc were fused into a bulb connected with a pump, and the electrode E placed above the arc and connected with an electrometer. This electrode was connected with the bulb by means of a piece of india-rubber tubing, through which it could be moved up or down, and so placed at various distances from the arc. At low pressures the difference in sign between the effects in hydrogen and oxygen disappeared, as wherever the electrode was placed it always got a charge of *positive* electricity, whether the gas was hydrogen or oxygen. At these low pressures no arc was visible between the terminals; the visible part of the discharge consisted of a glow which covered the terminals.

In another series of experiments a different method was employed: instead of testing the electrification by an electrode placed in the immediate neighbourhood of the arc, the gas from near the arc was sucked into a vat some two or three metres from the arc, and it was not until it had got into this vat that its electrification was tested. The arrangement used is represented in Fig. 11. The arc was formed in an inverted flask, and the air was sucked out from just over the arc. The air before entering the flask bubbled through sulphuric acid and passed through a tube filled with glass wool; another tube filled with glass wool was placed between the arc and the vat. The gas, after it had passed through the glass wool, carried strong positive electrification into the vat. This experi-

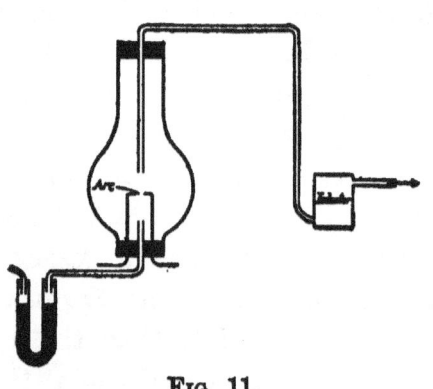

FIG. 11.

ment seems to show that the electrification observed near arcs, and probably therefore near incandescent solids, is not carried by metallic dust given off either from the electrodes or from the hot wire, but by the gas itself, though the gas carrying the electrification must, as the next experiment shows, be in an abnormal condition.

Though the electrification can pass through a plug of glass wool which is quite impervious to dust, yet the carriers of the electrification must be much more complex than the ordinary molecules of a gas, inasmuch as, though these molecules can readily pass through unglazed porcelain, the positive electrification carried by a gas seems quite unable to do so. This is shown by the experiment represented in Fig. 12.

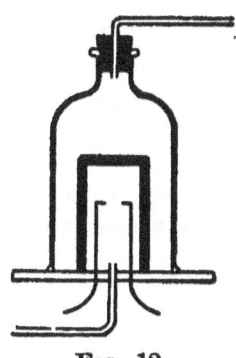

FIG. 12.

The arc passes between the terminals, which are placed inside a porous pot, such as is used for Daniell's cells. This pot is placed on plate and surrounded by a bell-jar, the joints between the plate and the pot, and the plate and the bell-jar, being carefully made air-tight. The air was sucked out of the bell-jar into the vat; but this air when examined showed no trace of electrification, though the coloured fumes of nitrous oxide in the bell-jar showed that air from the neighbourhood of the arc was being sucked into the vat.

An experiment was tried in which the air, before reaching the arc, was sucked through a porous pot, while only glass wool intervened between the arc and the vat; in this case there was strong positive electrification in the vat.

This experiment shows that the positive electrification is not carried by single molecules, but by some complex aggregation of them, — a result we are also led to by a study of the properties of electrification produced by chemical action in gases, and also by the effects of Röntgen rays.

The electrification produced by incandescent bodies and the arc can be explained by the same principles as those used to explain the electrification produced in some cases of combustion and chemical action.

Suppose that in consequence of the high temperature of the arc or incandescent metal, some of the molecules of the gas are dissociated into positively and negatively electrified atoms, and that some of these combine with the incandescent metal or with the terminals of the arc; then, in the compound formed, the atom of metal, being the electro-positive element, will have the positive charge; while the atom of gas which is the electro-negative element will be charged negatively. Thus to get this compound we take a negatively electrified atom from the gas and a positively electrified one from the metal, so that if to begin with there were as many positive as negative atoms both in the gas and the metal, the result of the formation of the compound would be to leave an excess of positive electrification in the gas and an excess of negative electrification in the metal. This is what we observe in the arc and incandescent solid for all metals when the gas surrounding them is anything but hydrogen. The negative electrification in hydrogen is anomalous, unless we could suppose that hydrogen was electropositive to platinum. I am inclined to attribute the small positive electrification observed in hydrogen when platinum is used to some secondary effect; it may perhaps be connected with the absorption of hydrogen by platinum, for if copper is used instead of platinum, the electrification in the hydrogen is positive.

At a white heat the temperature may be sufficiently high to dissociate the compound of the gas if formed, or to prevent its formation. In this case the molecules of the gas would be split up into atoms so that the gas would be a conductor; but inasmuch as there would be as many positively electrified atoms as negative, there would be no excess of either kind of electrification; thus the gas would conduct, but would not show any unilateral conductivity. Elster and Geitel found that unilateral conductivity disappeared when the incandescent metals were raised to a white heat.

It may be worth pointing out that, in consequence of the magnitude of the charges carried by the atoms an amount of electrification far exceeding anything observed near an incandescent metal could be produced by an amount of chemical combination far transcending the powers of chemical analysis to detect. Thus, if all the negative atoms were picked out and the positive ones left, from one millionth of a cubic centimetre of hydrogen at standard temperature and pressure, there would be enough electricity in the gas to raise a condenser whose capacity is 1000 cm. to the potential of about ten thousand volts.

We may mention some experiments made with the apparatus described which seem to corroborate this view. Thus, when sodium salts were placed in the arc, the electrification in the air sucked from the neighbourhood of the arc was positive, until so much sodium vapour was produced that the arc passed through this vapour instead of through the air; when this occurred the electrification in the gas sucked from the arc changed from *positive* to *negative*. This is what we should expect on the preceding view of the cause of the electrification, for as sodium is electro-positive to the material of which the terminals are made, if anything of the nature of combination were to take place, the positive sodium atoms would be taken and the negative ones left; there would thus be an excess of negative electrification in the sodium vapour. This method of sucking the air from the neighbourhood of the arc enables us to demonstrate the positive electrification in the neighbourhood of glowing metals in a way from which any effects that could be produced by metallic dust are excluded. We suck the air from the arc into the vat through a tube containing a plug of glass wool to stop the dust; we observe the amount of electrification carried by the gas into the vat due to the arc alone; we then introduce into the arc a piece of iron gauze, and find that as soon as this becomes red hot, there is a considerable increase of positive electrification in the vat.

Another experiment bearing very directly on the view that chemical action is the cause of the electrification observed near the arc and incandescent bodies is as follows : The arc was formed inside a vessel through which a stream of hydrogen was kept flowing, the hydrogen passing into the vat after leaving the vessel in which the arc was passing. The terminals used were copper, and were well oxidized before the commencement of the experiment. When the arc passed, the terminals got hot and the oxide was reduced ; while this process was going on, the hydrogen carried *negative* electrification into the vat. When, however, all the oxide had been reduced to copper and the arc passed between clean copper terminals, the *positive* electrification was carried into the vat. This is what we should expect on our theory, for while the reduction of the oxide is proceeding, the chemical action taking place is the combination of hydrogen with the oxygen of the oxide to form water; here the positively electrified atoms are taken for the compound, and the negatively electrified ones are left. There will thus be an excess of negative electrification in the hydrogen; when, however, the oxide is completely reduced, any combination must be between hydrogen and copper. As copper is electropositive to hydrogen, the negatively electrified atoms of hydrogen will be taken for this combination, and the positively electrified ones left; there will thus be an excess of positive electrification in the hydrogen.

For this experiment to be successful, the oxidized copper terminals should not be too thick; otherwise they will not get hot enough for the oxide on them to be completely reduced. We may compare this experiment with the one described on p. 84.

Conduction through Hot Gases.

It is in some cases more convenient to investigate whether or not a gas is a conductor of electricity than to determine whether or not it can receive an electric charge. Though a gas under normal conditions is a perfect insulator, yet under the action of various physical agencies it can become a conductor. One of the most interesting of the physical changes which can in some cases turn a gas from an insulator to a conductor is that of increase of temperature. Becquerel[1] found that air at a white heat would allow electricity to pass through it, even though the potential difference was only a few volts. This result was confirmed by Blondlot,[2] who found that air at a bright red heat allowed a current of electricity to pass through, even though the potential difference was as low as 1/1000 of a volt. He found that the conduction through the hot gas did not obey Ohm's law.

In some experiments which I made on this subject,[3] I found that hot gases conduct electricity with very different degrees of facility. Gases such as air, nitrogen, or hydrogen, which do not experience any chemical change when heated conduct

[1] Becquerel, Annales de Chimie et de Physique, [3] 39, p. 355, 1853.
[2] Blondlot, Comptes Rendus, 104, p. 283, 1887.
[3] J. J. Thomson, Phil. Mag., 5, 29, pp. 358, 441, 1890.

electricity when hot, but only to a very small extent, and in this case the conduction, as Blondlot supposed, appears to be convective. Gases, however, such as iodine, hydriodic acid gas, etc., which dissociate at high temperatures, conduct with very much greater facility.

In these experiments a large number of gases were examined, and, in every case where the hot gas possessed any considerable conductivity, I was able to detect by purely chemical means that chemical decomposition had been produced by the heat. Thus the heat makes the gas a conductor, because it produces a chemical change and dissociation, not because it raises the temperature. In this connection we must distinguish between two classes of dissociation. The first kind is when the molecule is split up into atoms; as, for example, in iodine, hydriodic acid gas, hydrochloric acid gas (where the chlorine, though not the hydrogen, remains partly dissociated). In all cases where dissociation of this kind occurs, the gas is a good conductor when hot. The second kind of dissociation consists in the splitting up of the molecules of the gas into simpler molecules, but not into atoms. Examples of this kind of dissociation are when a molecule of ammonia is split up into molecules of hydrogen and nitrogen, or when a molecule of steam splits up into molecules of hydrogen and oxygen. In these cases the gases only conduct on the very much lower scale of the non-dissociable gases.

The following gases only conduct slightly when heated: —

Air, nitrogen, carbonic acid, steam ammonia, vapour of sulphuric acid, vapour of nitric acid, sulphur in an atmosphere of nitrogen, sulphuretted hydrogen (in an atmospheric of nitrogen).

The gases in the following list all conduct well: iodine, bromine, chlorine, hydriodic acid gas, hydrobromic acid gas, hydrochloric acid gas, potassium iodide, sal-ammoniac, sodium chloride, potassium chloride.

Chemical analysis showed that all the gases in the second list were decomposed by the heat. Thus these experiments show that when a gas conducts on the scale of those in the second list, free atoms or something chemically equivalent to them must be present. The small amount of conductivity possessed by hot gases which are not decomposed by heat seems to be due to a convective discharge, carried, perhaps, by dust given off by the glowing electrodes; we must compare the results of Elster and Geitel's experiments on the electrification produced by glowing bodies; in consequence of this, gases in contact with glowing metals would acquire a certain amount of conductivity. The great difference between the conductivities of gases in the two lists shows, however, that there are other sources than the influence of the glowing electrodes by which gases can acquire electrical conductivity.

In all gases, however, the temperature of the electrode has a great effect upon the passage of electricity through the gas into which the electrodes dip. In the experiments described above I found it impossible to get a current of electricity to pass through the gas, however hot it might be, unless the electrodes were hot enough to glow. A current passing through the hot gas was immediately stopped by placing a large piece of cold platinum foil between the electrodes, even though a strong up-current of the hot gas was maintained to prevent the gas getting chilled by the cold foil. As soon as the foil began to glow, the current of electricity through the gas recommenced.

This is one among many instances of the difficulty with which electricity passes from a gas to cold metal.

I examined also the conductivities of several metallic vapours, including those of sodium, potassium, thallium, cadmium, bismuth, lead, aluminium, magnesium, tin, zinc, silver, and mercury. Of these the vapours of tin, mercury, and thallium hardly seemed to conduct at all; the vapours of the other metals conducted well, their conductivities being comparable with those of the dissociable gases.

Although mercury vapour does not conduct when heated, yet when exposed to Röntgen rays it conducts with great facility, and indeed to a greater extent than almost any other gas.

Conduction by Flames.

The gases which arise from flames conduct electricity, and they retain this power for some minutes after they have left the flame.[1] The laws of conduction through flame gases are in many respects analogous to those of conduction through gases exposed to the Röntgen rays. Thus Giese[2] (who was the first to suggest that electricity was carried through gases by the movement of oppositely charged ions) has shown that if the flame gases, on their way to the place where their conductivity is tested, are traversed by an electric current, the conductivity is very much diminished, just as in the case of the Röntgenized gas. Again, the current through the flame gases, when both electrodes are immersed in the gas, like that through the Röntgenized gas, does not increase so quickly as the electro-motive force. In some cases, when one of the electrodes is outside the flame gases, the current increases more rapidly than the electro-motive force; as, for example, in an experi-

[1] Giese, Wied. Ann., 17, p. 517, 1882.
[2] *Idem.*

ment due to Giese,[1] in which a long flame burned inside a metal cylinder, while between the cylinder and the flame there was an insulated metal ring connected with one pair of quadrants of an electrometer. When the potential difference between the flame and the outside cylinder was increased in the proportion of 1.7 to 1, the rate of flow of the electricity to the ring was increased in the proportion of 3.3 to 1. In this case, if there were originally an equal number of positive and negative ions in the flame, then under the electric field, if the cylinder is at a low potential and the flame at a high one, the ions will adjust themselves so that there is an excess of the positive ions at the surface of the flame, while the corresponding number of negative ions go to the earth through the flame. This excess of positive ions at the surface will be proportional to the potential difference; and it is at the surface that the potential gradient is greatest. Thus, not only does the potential gradient, and therefore the velocity of the ions increase as the potential gradient, but also the number of positive ions which carry the positive charge are greater when the potential difference is large than when it is small. Thus the rate at which the charge on the ring increases will increase more rapidly than the potential difference.

[1] Giese, Wied. Ann., 38, p. 403, 1889.

In Giese's experiments the rate at which negative electricity accumulated on the ring was always greater than that at which positive electricity accumulated, even though the potential difference was the same in magnitude; this indicates that under equal potential gradients the velocity of the negative ions is greater than that of the positive ones. No such difference has been observed in the case of gases traversed by the Röntgen rays.

Arrhenius[1] has shown that when solutions of salts of the metals of the alkali group are thrown into the flame by a spray producer, the conductivity of the flame is very greatly increased. The high conductivity of such flames is shown by the following table taken from Arrhenius's paper.[1] The current passed between two parallel platinum plates,

EMF in Clark's Cells.	KI	$\tfrac{1}{2}KI$	$\tfrac{1}{10}KI$	$\tfrac{1}{10^2}KI$	$\tfrac{1}{10^3}KI$	$\tfrac{1}{10^4}KI$	$\tfrac{1}{10^5}KI$
1	540	248	120	52.2	20	6.9	2.7
2	616	284	139	59.9	23.	7.9	2.9
5	734	360	174	75.8	29.1	10.1	3.7
10	1009	464	225	99.7	37.4	13.0	4.7
20	1340	616	298	130	49.6	17.2	6.3
40	1920	811	427	186	71.2	24.8	9.0

[1] Arrhenius, Wied. Ann., 42, p. 18, 1891.

about 1 cm. broad by 2.3 cm. high, separated by a distance of .56 cm. The unit of current is 10^{-8} amperes. The heading $\frac{1}{n}KI$ indicates that the solution sprayed into the flame is $\frac{1}{n}$th normal solution of potassium iodide.

These values are the difference between the currents when the given solutions are in the flame and when pure water is sprayed into the flame. The currents in the same units for the flame by itself are

$$E = 1 \quad 2 \quad 5 \quad 10 \quad 20 \quad 40$$
$$5.3 \quad 6.7 \quad 10.0 \quad 14.3 \quad 21.2 \quad 34$$

It is evident that in this case, as in that of the Röntgen rays, the current cannot exceed a certain value, for in the space between the electrodes only a certain amount of salt is introduced in unit time; the current therefore which is carried electrolytically by the salt cannot exceed that which would electrolyze this quantity in unit time. If we plot the curves representing the relation between C, the current, and E, we see that for large values of E the curve is not parallel to the axis along which E is measured, as it should be if the conduction were wholly electrolytic. The curve is, however, a straight line inclined to the axis of E, thus indicating a current which follows Ohm's law; it is probable that this part of the curve represents a current carried by solid particles in the flame. A detailed study of the relation between the current and the potential difference for conduction through flames would be likely to throw a good deal of light both on the processes going on in the flame and the nature of the conduction. In all conduction through flames the state and temperature of the electrode seems to have a very important influence on the result.

Effect of a Discharge in making a Gas a Conductor.

When an electric discharge passes through a gas, the gas is for the time being converted into a conductor; it conducts not merely in the direction of the original discharge, but in all directions, even those at right angles to the direction in which the primary discharge passed. The first to notice the change produced in a gas by the passage of a spark seems to have been Faraday,[1] who found that though it might be difficult to get the first spark to pass, yet when once this occurred, succeeding discharges followed quite readily.

Hittorf[2] showed that a few galvanic cells are able to send a current through a gas which is con-

[1] Faraday, Experimental Researches, § 392.
[2] Hittorf, Wied. Ann., 7, p. 614, 1879.

veying an electric discharge, though were the gas in its normal state, such an electro-motive force would be quite incapable of forcing electricity through the gas.

Schuster [1] describes a somewhat similar effect. A large discharge tube containing air at a low pressure was divided into two compartments by a metal plate with openings round the perimeter; these openings allowed the gas in one compartment to flow into the other, while the metal screened off from one compartment any electrical action occurring in the other. Schuster found that if a vigorous discharge passed in one compartment, an electromotive force of about one quarter of a volt was sufficient to send a current through the gas in the other.

Both cathode and Lenard rays make gas through which they are travelling into a conductor.

I have found that a gas (at low pressure) inside a closed metallic vessel, connected with the earth and placed inside a tube in which an electric discharge is passing, becomes under some circumstances a conductor.

The arrangement used was that shown in the following diagram (Fig. 13). A and B are the electrodes of the discharge tube, A being a disk and B a wire. At the further end of this tube

[1] Schuster, Proceeding Roy. Soc., 42, p. 371, 1887.

another long tube was fused on; in this was placed a long cylinder made (except for the end facing the discharge) of brass about 1.5 mm. thick. This was pierced by a small hole at the further extremity of the tube, so as to make the pressure of the gas inside the cylinder at the same pressure as that outside; for the end of the cylinder facing the discharge, various metals were used, — very thin aluminium foil, thick aluminium foil about .1

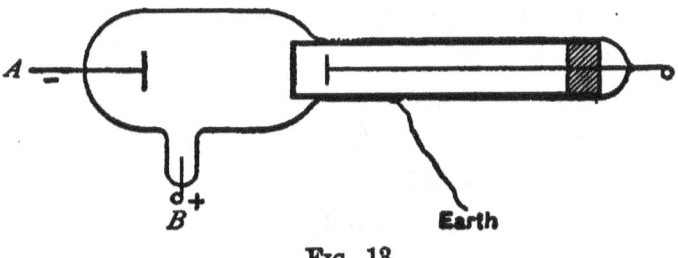

FIG. 18.

mm. thick, sheet brass about .1 mm. thick, and brass about 1.5 mm. thick. This cylinder was connected with the earth. Inside the cylinder and carefully insulated from it, was a rod carrying a disk which was placed parallel and close up to the flat end of the cylinder. This disk was connected with an electrometer, the connecting wire being surrounded by earth-connected conductors to screen it from electrostatic influence. The disk was carefully insulated from the outer cylinder. When the pressure in the tube was reduced just enough to allow the discharge to begin to pass, then although the discharge was passing, the disk did not receive any charge, neither if it were charged to begin with, did it lose its charge; the gas inside the cylinder is thus still in its normal state. When, however, the tube was still further exhausted and the cathode began to be covered with a velvety glow, but long before any phosphorescence appeared on the walls of the tube, a faint glow began to flicker over the end of the earth-connected cylinder, and if this were the one with the end made of the thin aluminium foil, the disk received a charge of positive electricity; and if it were charged up to begin with, it lost its charge. The rate of leak was more rapid when the original charge was negative than when it was positive. The gas inside the cylinder has thus been put in a peculiar state by the discharge going on outside, though protected from it by a cylinder connected with the earth. Let us for the present confine our attention to the case of the cylinder with the end made of thin aluminium foil. As the pressure is still further reduced and the dark space round the negative electrode reaches an appreciable thickness, the positive charge on the disk diminishes and finally vanishes; but though the disk receives no charge, if it is charged up it will not retain its charge. There is a rapid leak which is now about equally rapid, whether the disk is charged positively or negatively. When the pressure is still further reduced and the phosphorescence appears on the walls of the tube, the disk acquires a negative charge, and the leak from the disk is more rapid when the original charge is positive than when it is negative. When the pressure is so low that the cathode rays fall upon the ends of the cylinder, there is again a negative charge on the disk and a leak from it; but the rate of leak diminishes very greatly if the cathode rays are deflected by a magnet and so prevented from reaching the ends of the cylinder. The leak for both positive and negative charges was diminished when the cathode rays were deflected from the cylinder. If at a low pressure the disk A was made the anode and the wire B the cathode, there was no appreciable charge, but still a leak. If the cylinder itself were made the anode and still kept in connection with the earth, the disk received a negative charge; when the cylinder was cathode at certain pressures the disk got a positive charge, and at all pressures the rate at which positive electricity escaped from the disk was greater when the cylinder was anode than when it was cathode. The preceding experiments were made with an induction coil. The experiments at the higher pressures were repeated, using instead of the induction coil a battery of 1000 small storage cells: results quite similar in character to those previously described were obtained; the experiments at high exhaustion could not, however, be tried with these cells, as their electro-motive force was not large enough to produce a discharge at pressures low enough to allow the cathode rays to travel as far as the end of the cylinder.

Experiments were next tried with the cylinders with different ends, to see what effect the thickness and materials of which the ends were made had upon the phenomena. To do this, four similar and equal tubes were prepared, each like the one in Fig. 13. In one of these the cylinder with the thin aluminium end was placed; in the second, the one with the thick aluminium; in the third, the one with the brass, .1 mm. in thickness, and in the fourth, the one with the thick brass (about 1.5 mm. in thickness). These were connected together, and with the pump, so that the pressures in the four tubes were always the same. Of these cylinders the one with the thinnest aluminium end was the one where the charge on the disk first appeared, for the gas in this cylinder leaked when that in all the others insulated perfectly; when, however, the pressure was so low that the disk inside the thin aluminium cylinder got a negative charge, the disk, inside the thick aluminium and the thin brass cylinders also got a slight negative charge, though not so large a one as that inside the thin aluminium cylinder. At this stage too the insulation of the gas inside these cylinders broke down; the leak when the disks were positively electrified was faster than that when they were negatively electrified: at this stage the disk inside the thick brass cylinder did not receive a charge uor did the insulation of the gas break down. On reducing the pressure still further, however, a stage was reached when the disks inside this cylinder, like the other disks, received a negative charge, and when the gas inside it leaked. At this stage the cathode rays reached the cylinders, and the rate of leak for the thick cylinders as well as for the thin was diminished by turning with a magnet the rays from the end of the cylinder. The effect in the thick cylinder was always least; that in the thinnest aluminium one always the greatest. The effects were specially marked where the earth-connected cylinder was used as the anode, and the disk A as the cathode.

These experiments show that the gas (at a low pressure) inside a closed metallic vessel connected with the earth becomes a conductor when the vessel is placed inside a highly exhausted tube through which an electric discharge is passing; that the conductivity increases as the thickness of the walls of the metallic vessel diminishes, but is appreciable when the walls are 1/16 of an inch thick; and that the incidence of cathode rays on the vessel increase very much the conductivity of the gas.

We now pass on to the consideration of the question how the gas inside the cylinders gets affected by the discharge in the gas outside it. In the first place, are the phenomena consistent with the idea that rays of some kind are produced by the discharge and travel right through the walls of the cylinder? We know that cathode, Lenard, and Röntgen rays render every gas through which they pass a conductor of electricity, and they have the property of passing through thin films of metal which are quite opaque to ordinary light. So far they seem to have properties which would explain the effects we have just described; but a closer examination shows that this explanation is not feasible. In the first place, the effects we have described began at pressures far too high for cathode or Röntgen rays to be produced, again these rays would not pass through brass 1/16 of an inch thick, and lastly when the cathode rays fall on a conductor they give it a negative charge; while the disk in our experiments gets, when the pressure is high, a positive charge.

It seems to me much more probable that the charge on the disk and the conductivity of the gas inside the cylinder are caused by an electric discharge between the cylinder and the disk. This outer cylinder is, however, kept connected with the earth; and if the electrical conditions in the discharge tube outside the cylinder were steady, the cylinder would be at zero potential, or, if an electric current were continually flowing from the outer cylinder to the earth, would differ from it at most by the small potential difference corresponding to this current, — a potential difference which would be quite insufficient to produce a discharge between the cylinder and the disk. Thus, if the effects we have described are due to discharges taking place from the cylinder, the electric discharge through the gas at low pressure outside the cylinder cannot be continuous.

If, however, this discharge is intermittent we might, I think, expect that even though the cylinder were kept connected with the earth, its potential would be subject to large fluctuations, and might attain to values which would make discharge between the cylinder and the disk possible.

For suppose that the region just outside the cylinder were very suddenly to receive a charge of electricity, let us say negative, then in order to neutralize in the inside of the cylinder the electric intensity due to this charge, the cylinder must get a positive charge and negative electricity run to the earth. This will require a finite, although very short time; but until it is completed the potential of the part of the cylinder near the charge will be lower than that of the earth. This difference of potential, though only transient, may last sufficiently long to produce an electric discharge between the cylinder and the disk, and thus to give the disk a negative charge, while the passage of the discharge will turn the intervening gas into a conductor.

Thus we see that if a negative charge of electricity is suddenly brought up to one side of a plate, the other side will for a short time act like a cathode; while it will act as an anode if a positive charge is suddenly brought up to the outside of the cylinder.

The laws of diffusion of electric intensity through a mass of metal are the same as those of temperature in a conductor of heat; hence if an electric intensity is suddenly started at the surface of the metal, it will diffuse through the metal, and the maximum electric intensity at any point in the metal will vary as some inverse power of the distance of the point from the surface of the metal. The power will depend upon the initial distribution of the electric intensity. The thinner the sides of the metal cylinder, the greater will be the electric intensity in the inside, due to the sudden appearance of a charge on the outside.

CONDUCTIVITY OF GASES

The experiments described above show that if this explanation is the true one, then at the highest pressure at which the effects were observed, the charge suddenly brought up to the cylinder must have been a positive one, and this was not large enough to be appreciable in any of the cylinders except the one whose end was very thin aluminium foil. When the pressure was lower, and especially when the cathode rays began to reach the cylinder, the charge suddenly brought up to the cylinder was negative, and was now great enough to produce an effect even through one-sixteenth of an inch of brass.

The experiments showed, too, that this negative charge was greatly diminished when the cathode rays were deflected from the cylinder by a magnet; this is quite in accordance with well-known properties of these rays, for we shall see later that the cathode rays carry charges of negative electricity.

As these effects are obtained when the discharge is produced by a battery of storage cells, as well as by an induction coil, we must, if the preceding explanation is to apply, suppose that the discharge in the gas is not uniformly continuous, even when produced by a large battery of galvanic cells.

The question of the continuity or discontinuity of the electric discharge through a gas at a low pressure is one that has attracted a considerable amount of attention, but on which physicists have not arrived at any very general agreement.

In the following remarks we shall confine our attention to the evidence with respect to the continuity and steadiness of the discharge produced by a large battery, as the conditions for continuity are more favourable when the electric discharge is produced in this way than when it is produced either by an induction coil or an electric machine.

The first physicist to send an electric discharge through a gas by means of a large number of galvanic cells was Gassiot. He found that the electric discharge he produced in this way was evidently intermittent; for when it was examined by means of a rapidly rotating mirror, the image of the discharge in the mirror, instead of being a continuous ribbon of light, as it would have been had the discharge been continuous, was broken up into a series of bright and dark stripes. Gassiot's battery had, however, a very high internal resistance. Hittorf showed that a discharge which could not be proved to be intermittent by the use of the rotating mirror could be obtained by using a battery with a sufficiently small internal resistance, and that by interposing in the circuit a high resistance, the discharge could be changed into one that was plainly intermittent.

When the high resistance is in the circuit, the tube, or better a telephone connected up with the tube, gives out a note whose pitch indicates the rate at which the discharges succeed each other. As the resistance diminishes, the pitch of the note gets higher and higher; but if the resistance is gradually diminished until the discharge passes into the state in which it can no longer be resolved into separate discharges by the rotating mirror, the transition from the one discharge to the other does not take place gradually. When the resistance reaches a certain value, the note suddenly ceases, and then no further evidence of the discontinuity of the discharge can be obtained. This is made more striking when the electrodes of the tube are connected with the plates of a large condenser. From this abruptness in the change in the discharge, Hittorf concluded that the discharge with the low resistance in the circuit was as continuous as the passage of a current through a metal or an electrolyte.

It is remarkable how the various gases differ in the ease with which the discharge through them loses the characteristics of discontinuity. In some experiments made by Mr. Capstick in the Cavendish Laboratory it was found perfectly easy to get the discharge to pass through the simple gases, such as hydrogen or nitrogen, without producing any sound in a telephone connected with the tube; with compound gases, however, the result was very different. Thus it was found exceedingly difficult to get the conditions such that the telephone was quiet when the discharge passed through water vapour, and impossible to do so when it passed through ammonia.

Hertz[1] made a series of experiments to see if he could detect, using much more delicate means than the revolving mirror, any intermittence in the discharge. Thus he inserted in series with the tube a galvanometer and an electro-dynamometer, and compared the readings of the two. He then replaced the tube and large battery by a Daniells' cell, and inserted resistances until the deflection of the galvanometer was the same as before; he found that when this was the case the deflection of the dynamometer was the same also. But if the current through the exhausted tube had been intermittent, then if the galvanometer deflection was the same as for the steady current, the deflection of the dynamometer ought to have been greater in the first case than in the second.

Another method used by Hertz was to connect the gold leaves of an electroscope by a short wire with the cathode, while the metallic case of the electroscope was connected with the cathode through an exceedingly large resistance. The

[1] Hertz, Wied. Ann., 19, p. 782, 1883.

idea was that if any fluctuation took place in the potential of the cathode, the gold leaves of the electroscope, being connected by a short wire with the cathode, will be able to follow these fluctuations in the potential, while the great resistance between the cathode and the case of the electroscope will prevent the latter doing so. Thus a difference of potential would generally exist between the gold leaves and the case of the instrument, which might be sufficient to produce a repulsion of the gold leaves. No repulsion, however, was observed. These experiments of Hertz avowedly prove only at the most that the rate of intermittence, if not infinite, is exceedingly large.

We shall do well, I think, to distinguish between the current through the tube and that through the leads. The current through the former may be discontinuous, even though that through the latter is continuous. For since the current through the gas does not obey the same laws as that through a metallic conductor, the current across a section of the discharge tube need not at any specified instant be the same as that across a section of one of the leads. The average current must be the same in the two cases, but only the average current, and not that at any particular instant. To quote an illustration given by Spottiswoode and Moulton, the discharge tube may act like the air vessel of a fire-engine; all the electricity that comes in goes out again, but no longer with the same pulsations. The tube may sometimes contain more and sometimes less free electricity, and may act as an expansible vessel would act if it formed part of the path of an incompressible fluid.

Again, with respect to the very high rate of intermittence, we must remember that we should get what, as far as its effects went, would be equivalent to this, if the discharge, instead of coming simultaneously off the whole electrode, came off from one place at one time and another place at another time, even though the discharges at any one place followed each other comparatively slowly. An illustration may make this clearer. Let us compare the discharge from an electrode to the firing of a battalion of 5000 men; and suppose each man fires 3 rounds a minute, then if they fire in volleys, the firing will have the period of 20 seconds. If, however, they do not fire in volleys but each man when he likes, an observer would have to be able to distinguish intervals of about 1/250 of a second if he wished to detect intermittence. There are many circumstances which favour the view that at very low pressures the discharge comes off independently from different parts of the same electrode.

So far as I know there is no phenomenon inconsistent with the view that variations are continually occurring in the electrical state of a tube through which a discharge is passing, and this variation would account for the phenomena described above.

Electrolysis in Gases.

In the preceding account of the conductivity conferred on gases by light, Röntgen rays, and heat, we have seen that the phenomena can be explained on the supposition that the electricity gets through the gas by the movement of oppositely charged ions through the gas, the process being similar to that by which electricity is carried through an electrolyte. The passage of sparks through a gas furnishes us with additional evidence in favour of this view.

The first direct evidence that the conduction of electricity through gases was effected by electrolysis was given by Perrot[1] in a remarkable investigation he made on the passage of the discharge through water vapour. The apparatus used by Perrot in his experiments is represented in Fig. 14. The sparks passed between two platinum wires sealed into glass tubes, *cfg*, *dfd*, which they did not touch except at the places where they were sealed in. The open ends, *c, d*, of these tubes were about 2 mm. apart, and the wires terminated inside the tubes at a distance of about 2 mm. from the

[1] Perrot, Annales de Chimie et de Physique [361], p. 161, 1861.

ends. The other ends of these tubes were inserted under test tubes, *ee*, in which the gases which passed up the tubes were collected. The air was exhausted from the vessel *A*, and the water vapour through which the discharge passed was obtained by heating the water in the vessel to about 90° C. Special precautions were taken to free this water from any dissolved gas. The stream of vapour

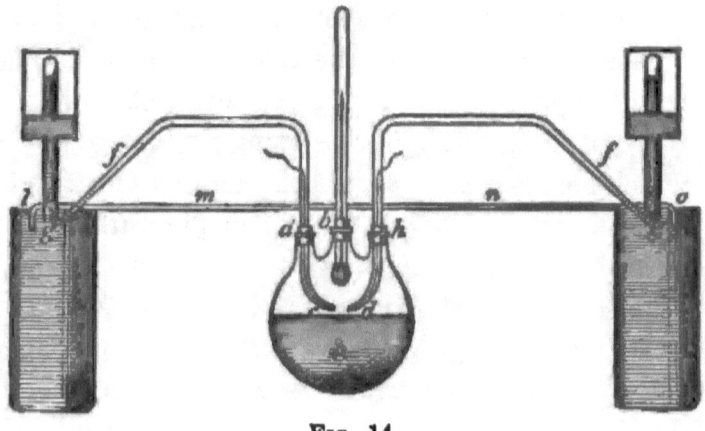

Fig. 14.

arising from this water drove up the tubes the gases produced by the passage of the spark. Part of these gases was produced along the length of the spark; but in this case the hydrogen and oxygen would be in chemically equivalent proportions. Part of the gases driven up the tubes would, however, be liberated at the electrodes; and it is this part only that we could expect to bear any simple relation to the quantity of electricity which had passed through the gas.

When the sparking had ceased, the gases which had collected in the test tubes, ee, were analyzed; first, they were exploded by sending a strong spark through them. This at once got rid of the hydrogen and oxygen, which were in chemically equivalent proportions, and thus got rid of the gas produced along the length of the spark. After the explosion the gases left in the tubes were the excesses of hydrogen and oxygen, together with a small quantity of nitrogen, due to a little air which had leaked into the vessel in the course of the experiments, or which had been absorbed by the water. The results of these analyses showed that there was always an excess of oxygen in the test tube in connection with the positive electrode, and an excess of hydrogen in the test tube connected with the negative electrode, and also that the amounts of hydrogen and oxygen in the respective tubes were very nearly chemically equivalent to the amount of copper deposited from a solution of copper sulphate in a voltameter placed in series with the discharge tube.

These results seemed to me so important that I repeated them, making some changes in the apparatus to meet some objections to which it appeared to me Perrot's form was liable.

The method used is shown in Fig. 15.

H is a glass bulb, 1.5 to 2 litres in volume, containing the water which supplies the steam. A tube, *L*, about 0.75 cm. in diameter and 35 cm. long, is joined on to this.

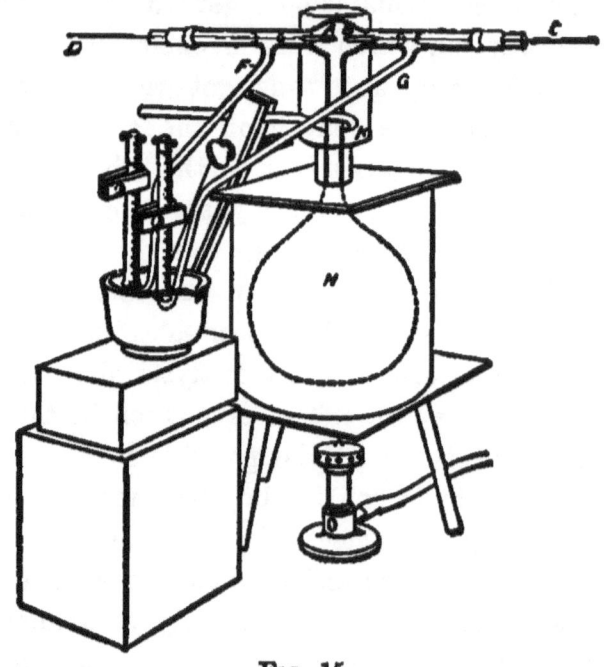

FIG. 15.

The top of the tube *L* is fused on to the horizontal discharge tube *CD*; this tube is blown out into a bulb in the region where the sparks pass, so that when long sparks are used they may not fly to the sides of the tube. The top of the tube *L*,

ELECTROLYSIS IN GASES

near its junction with CD, is encircled by a ring burner K, and this part of the tube is surrounded by an asbestos case; by these means the steam may be superheated to a temperature of 140° to 150° C.

The details of the electrodes between which the sparks pass are shown in Fig. 16. For the metal parts a, b, it is necessary to use some metal which is not oxidized by the steam, as a very small amount of oxidation would be sufficient to render the results nugatory.

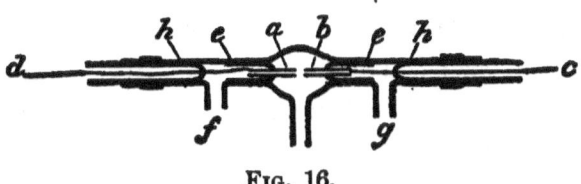

Fig. 16.

The glass tubes e, e, stop short of the places f, g, where the delivery tubes join the discharge tube. The discharge tube is closed at the ends by two pieces of tube, h, h, which have their ends inside the tube fused up; wires connected to the electrodes c, d, are fused through the closed ends of these tubes. It is desirable that the closed ends of the tubes h, h, should come up as near as possible to the exits f, g, as air is very apt to remain in the tube if there are any places through which the steam does not rush. The tubes h, h, may either be fused on to the spark tube or fastened to it by rubber tubing.

The delivery tubes F, G (Fig. 15), are fused on to the discharge tube at f, g (Fig. 16). These tubes are about 5 cm. in diameter and terminate in narrow openings. It is essential that the steam and the mixed gases should escape through the tubes F, G, at approximately the same rate; to ensure this, the narrow extremities of these tubes should be equal both in length and width. This was attained by drawing out a piece of tubing which was originally of the same diameter as F, G, and then cutting it at the middle of the narrow part; the two halves were then either fused or fastened by rubber tubing to F, G. The narrow ends of F, G, are turned up and placed under mercury in the vessel M (Fig. 15). Over these ends, graduated eudiometer tubes are placed; these are filled with mercury at the beginning of the experiment, but the mercury soon gets displaced by the water produced by the condensation of the steam rushing through the tubes.

The heat produced by this condensation serves a useful purpose; it raises the temperature of the water in the eudiometer tubes over which the gases are collected to over 80° C., and thus, since hot water absorbs oxygen but not hydrogen much less readily than cold, diminishes the disturbing effect due to the greater absorption of the oxygen than of the hydrogen by the water over which the gases are collected.

The effect produced by electrification on the condensation of a jet of steam is shown in a very striking way by this apparatus. When the delivery tubes are open to the air, the steam, after escaping from the nozzles, goes some inches before it condenses sufficiently to form a cloud; as soon, however, as the coil is turned on and the sparks pass, brownish clouds reaching right down to the nozzles are at once formed. The cloud is denser in the steam which has gone past the negative electrode than in that which has gone past the positive.

Great care was taken to get rid of any air which may have been in the apparatus or dissolved in the water. The sparks were produced by a large induction coil. After it had been ascertained that all the air had been expelled from the vessel and from the water, and that the rates of flow through the delivery tubes were approximately equal, the eudiometer tubes filled with mercury were placed over the ends of the delivery tubes, a water voltameter was placed in series with the steam tube, and the coil set in action.

The steam which went up the eudiometer tube condensed into hot water, which soon displaced the mercury; the mixture of oxygen and hydrogen produced by the spark went up the eudiometer tubes and was collected over this hot water, and exploded at short intervals of time by the sparks from a Wimshurst machine. The gases did not disappear entirely when the sparks passed; a small fraction of the volume remained over after each explosion, and the volume which remained was greater in one tube than in the other. The residual gas which had the greatest volume was found on analysis to be hydrogen; the other was oxygen. When a sufficient quantity of the residual gases had been collected, they were analyzed. The result of the analysis was that when the sparks were not too long, the residual gas in one tube was pure hydrogen, that in the other pure oxygen. If any other gases were present, their volume was too small to be detected by my analyses. When the sparks were very long, there was always some other gas (nitrogen?) present, sometimes in considerable quantity.

The results obtained by the preceding method varied greatly in their character with the length of the spark; I shall therefore consider them under the heads, "short sparks," "medium sparks," and "long sparks."

The lengths at which a spark changes from "short" to "medium," and then again to "long," depend on the intensity of the current passing through the steam, and therefore upon the size of the induction coil and the battery power used to drive it. The limits of "short," "medium," and "long sparks" given below must therefore be understood to have reference to the particular coil and current used in these experiments. With a larger coil and current these limits would expand; with a smaller one they would contract.

Short Sparks.

I shall begin by describing the experiments with short sparks; that is, sparks from 1.5 to 4 mm. long. Here the appearance of the spark shows all the characteristics of the "arc" discharge.

The discharge passes as a thickish column with ill-defined edges, and when placed in a wind it is blown out to a broad flame-like appearance.

For these short sparks, or "arcs," as I prefer to call them, two very important laws were found to be true: —

1. That within the limit of error of the experiments the volumes of the excesses of hydrogen in the one tube, and of oxygen in the other, which remain after the explosion of the mixed gases, are, respectively, equal to the volumes of the hydrogen and oxygen liberated in the water voltameter placed in series with the steam tube.

2. The excess of hydrogen appears in the tube which is in connection with the *positive* electrode, the excess of oxygen in the tube which is in connection with the *negative* electrode.

Medium Sparks.

When the spark length is greater than 4 mm., the first of the preceding results ceases to hold. The second of these, that the hydrogen comes off at the positive electrode, remains true until the sparks are some 11 mm. long; but, instead of the hydrogen from the steam being equal to that from the water, it is, when the increase in the spark length is not too large, considerably greater.

This increase in the ratio of the hydrogen from the steam to that from the voltameter does not continue when the length of the spark is still further increased. When the spark length has got to 8 mm., this ratio begins to fall off very rapidly as the spark length increases, and we soon reach a spark length at which it seems almost a matter of chance whether hydrogen or oxygen appears in the collecting tube connected with the positive electrode.

Long Sparks.

When the spark length is increased beyond the critical value, the excess of hydrogen, instead of appearing as with shorter sparks at the positive electrode, changes over to the *negative;* the excess of oxygen at the same time going over from the negative to the positive electrode. Thus the gases, when the spark length is greater than its critical value, appear at the same terminals as they do when released from an ordinary electrolyte, instead of at the opposite terminals, as they do when the sparks are shorter.

With regard to the quantity of hydrogen liberated from the steam in comparison with that set free in the voltameter, I find that when the spark length is a few millimetres greater than the critical length, the amount of hydrogen from the steam is very approximately the same as that in the voltameter. The following table contains a few measurements on this point: —

Spark length.	Hydrogen from steam.	Hydrogen from voltameter.
10 mm.	0.7 c.c.	0.8 c.c.
12* "	0.75 "	0.9 "
14 "	0.8 "	1.1 "

When the sparks are longer than 14 mm., the amount of hydrogen from the steam was no longer equal to that from the voltameter. The results, however, were irregular, and, as mentioned before, there was a considerable quantity of nitrogen (?) mixed with the hydrogen and oxygen.

The preceding results show that in the electrolysis of steam, as in that of water, there is a very close connection between the amounts of hydrogen and oxygen liberated at the electrodes and the quantity of electricity which has passed through the steam, and that this relation for certain lengths of sparks is the same in steam as in electrolytes. There is, however, this remarkable difference between the electrolysis of steam and that of water, that whereas in the case of water the hydrogen always comes off at the negative, the oxygen at the positive, electrode, in the case of steam the hydrogen and oxygen come off sometimes at one terminal, sometimes at the other, according to the nature of the spark.

The preceding method is only applicable in a very limited number of cases, as it is essential for the success of the method that the undecomposed gas, as well as the decomposed gases which are in chemically equivalent proportion, should be eliminated. This is possible when the gas is steam, which condenses after leaving the discharge tube, while the oxygen and hydrogen produced by the decomposition of the steam can be exploded. There are, however, hardly any other gases for which these conditions are fulfilled. When qualitative results only are required, other methods are available which are of very much more general application. Thus, if we take a discharge tube made of a piece of thermometer tubing of fine bore, with two small bulbs at the ends for the electrodes, and fill this with hydrochloric acid gas at a low pressure, the discharge is at first of a uniform greenish gray. However, after it has passed for some time, the colour changes, becoming green in the neighbourhood of the positive terminal and red at the negative. Now, when a discharge passes through chlorine, its colour is green; while when it passes through

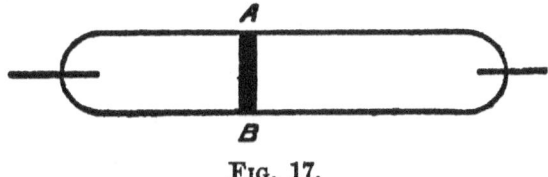

Fig. 17.

hydrogen it is red. Thus the experiment shows that when the discharge passes through hydrochloric acid gas the chlorine accumulates at the positive terminal, the hydrogen at the negative.

Another method which is capable of more general application is to use a tube like that shown in Fig. 17. This differs from an ordinary discharge tube merely in having a flat metal plate, AB, fastened across the tube. When the discharge passes through the tube, one side of the plate acts as a positive, the other as a negative, electrode. The tube is mounted on a stand which an observer at a spectroscope can move by means of a lever so as to bring one side or other of the plate opposite the slit of the spectroscope. If the tube is filled with a compound gas, then when the pressure is within certain limits on the side of the plate which acts as the positive electrode, the lines of the electro-negative constituent of the gas are strong, those of the electro-positive element either weak or absent; while on the side which acts as the negative electrode these conditions are reversed, — the lines of the electro-positive constituent are strong, those of the electro-negative one weak. Thus, if the tube is filled with hydrochloric acid gas, the hydrogen lines are much brighter on the negative side of the plate than on the positive, while the chlorine lines are brighter on the positive side than on the negative. If the tube is filled with ammonia, the hydrogen lines are bright on the negative side of the plate, absent from the positive side; on the positive side of the plate there is the positive pole spectrum of nitrogen, on the negative side the negative pole spectrum of nitrogen, and the hydrogen spectrum.

Transport of one Gas through Another.

This can be shown by using a tube like that shown in Fig. 18. It is made of very fine bore thermometer tube. The extremities of the tube in which the electrodes are fused are bent down so as

ELECTROLYSIS IN GASES 135

to be parallel to each other and so near together that a slight motion of the tube suffices to bring either electrode in front of the slit of the spectroscope. A side tube provided with two taps, A and B, is attached to the main tube; in the space between these taps a small quantity of gas can be imprisoned, and, by opening the tap A, let into the side tube. If the tube is filled with hydrogen at a low pressure, and a little chlorine let into the tube, then, after the discharge has been running

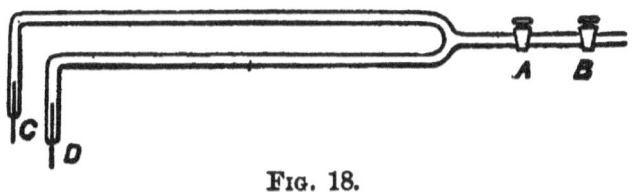

Fig. 18.

for a short time, the chlorine spectrum is found to be bright at the positive electrode, while no trace of it can be detected at the negative; on reversing the coil, the first effect is that the chlorine spectrum flashes out with increased brilliancy at the old positive electrode (which is now the negative). This, however, only lasts for a second or two; the chlorine spectrum rapidly fades away, and for a time is not visible at either electrode. It soon, however, reappears at the new positive electrode, the chlorine having thus been transferred from one electrode to the other. This process can be repeated any number of times.

If a little sodium vapour is introduced into a tube filled with air, the tube being placed in a sand bath, to prevent the sodium vapour condensing on the sides of the tube, the sodium travels to the negative electrode. Thus the sodium travels in the opposite direction to the chlorine. These experiments suggest that the separation of two gases, A and B, by the discharge is due to the decomposition by the discharge of a chemical compound formed of A and B, in which the A atoms have a charge of electricity of one sign, the B atoms a charge of electricity of the opposite sign, and that these charged atoms, under the influence of the electro-motive force in the tube, travel in opposite directions.

CATHODE RAYS

Some of the most interesting and important effects connected with the passage of electricity through gases are those associated with what are called cathode rays.

Plücker[1] seems to have been the first to observe, in 1859, the greenish phosphorescence on the walls of the glass tube near the negative electrode. He ascribed this to rays which diverge from the negative electrode, travel to the walls of the tube, and from thence back again to the negative electrode; he found that these rays were affected by a magnet in a different way from that in which the discharge ending at the positive electrode was affected.

Hittorf,[2] who greatly extended our knowledge of the subject, found that the agent producing the phosphorescence was intercepted by a solid or liquid, whether conductor or insulator, placed between the cathode and the walls of the tube. As

---

[1] Plücker, Pogg. Ann., 107, p. 77, 1859; 116, p. 45, 1862.
[2] Hittorf, Pogg. Ann., 136, p. 8, 1869.

the result of his researches, he came to the conclusion that the parts of the surface of the negative electrode are the starting-points of a motion which in a gas spreads uniformly like rays, and coincides in this respect with wave motion.

Goldstein,[1] who was the next to write about the subject, confirmed Hittorf's observation, that a body placed between the walls of the tube and a pointed cathode throws a well-defined shadow. He found, too, that well-defined though not very sharp shadows of objects may be obtained when the negative electrode is a large surface and placed at a small distance from the object. This shows that the rays which produce the phosphorescence of the glass must be emitted almost normally, and not like light, in all directions, for if the negative surface had been luminous, it would hardly throw a visible shadow of a small body placed near it. Goldstein, who introduced the name "Kathodenstrahlen," regards them as waves in the ether, for whose propagation the gas was not necessary.

Crookes,[2] who next investigated the subject, and who made some very striking discoveries which we shall describe in detail when we discuss the

---

[1] Goldstein, Monat der Berl. Akad., p. 284, 1876.
[2] Crookes, Phil. Trans., 1879, part i., p. 135; part ii., 1879, p. 587.

properties of the rays, took a different view: he regarded these rays as consisting of negatively electrified particles, projected at right angles to the cathode with great velocity, causing phosphorescence and heat by their impact with the walls of the tube, and being deflected by a magnet, since a magnet exerts on a moving charged body a force at right angles both to the direction of motion and to the magnetic force.

Properties of the Cathode Rays.

The cathode rays, when they strike against the glass walls of the discharge tube, produce phosphorescence. The colour of the phosphorescent light depends on the kind of glass. Thus the phosphorescent light from soda glass is a greenish yellow, while that from lead glass is blue. The cathode rays also produce luminosity in the gas through which they pass. An experiment to be described later on, however, shows that the luminosity in the gas is not always proportional to the brightness of the phosphorescence on the glass. Some of the rays from the cathode which produce very bright phosphorescence when they fall on the glass produce very little luminosity in the gas through which they pass.

A very large number of substances become phosphorescent when exposed to the cathode rays. The spectrum of the phosphorescent light given out by these bodies is usually a continuous one. Mr. Crookes, however, has found that when the cathode rays fall on some of the rare earths, such as yttrium, the substance gives out a spectrum with bright bands; and he has founded on this observation a spectroscopic method which is of the greatest importance in the study of the rare earths. An account of this method is given in Mr. Crookes's papers on Radiant Matter Spectroscopy, Phil. Trans., 1883, pt. iii., and 1885, pt. ii.

In most cases the light given out by bodies phosphorescing under the cathode rays is not polarized. Crystals of tinstone, however, when phosphorescing, give out polarized light,[1] the plane of polarization of the emitted light being parallel to the axis of the crystal; while crystals of hyacinth give out a phosphorescent light, where the ordinary ray is a different colour from the extraordinary.[2] In one specimen the ordinary ray was pale pink, the extraordinary lavender blue; in a second specimen the ordinary ray was pale blue, the extraordinary deep violet; while in a third crystal the ordinary ray was yellow, the extraor-

---

[1] Crookes, Phil. Trans., part ii., 1879, p. 661.
[2] *Ibid.*, p. 662.

dinary deep violet blue. In some cases the impact of the cathode rays on a substance produce a very apparent change in the colour, showing that they have changed the constitution of the substance. This is the case with the chlorides of silver, mercury, and lead, and, as Goldstein[1] has shown, with the haloid salts of the alkaline metals. Wiedemann and Schmidt[2] have traced the change in colour in the last case to the formation of a subchloride. In these cases the change in the chemical composition is rendered evident by the alteration in the colour. There are other cases, however, where, though these colour changes do not occur, there are other indications that a change in the composition of the substance has taken place. Such indications are afforded by the phenomena called by E. Wiedemann Thermo-luminescence. Bodies which have been exposed to the cathode rays are found to possess for some time the power of becoming luminous when their temperature is raised to a point far below that at which they become luminous in their normal state. They retain this property for weeks and even months after exposure to the cathode rays, and the effects seem to indicate that some chemical change has taken place, and that the gradual transformation

[1] Wied. Ann., 54, p. 371, 1895.
[2] *Ibid.*, 54, p. 618, 1895.

of the substance back to its original state is accompanied by the emission of light; or that in this state they can under the influence of dark heat radiation emit luminous radiation. The gas in the bulb in which the cathode rays are produced does not seem to have any effect on the colour of the phosphorescent light, or on the changes of colour in the haloid salts of the alkaline metals. Phosphorescence under cathode rays is strikingly shown by substances which belong to the class of bodies called by Van't Hoff[1] solid solutions. These are formed when two salts, one greatly in excess of the other, are simultaneously precipitated from a solution; under these circumstances the connection between the substances seems more intimate than in an ordinary mechanical mixture, and a substance is formed which is very phosphorescent when struck by cathode rays.

The effect produced on the phosphorescence by the addition of a trace of a second substance is shown by the results given in the following table, which is taken from a paper by Wiedemann and Schmidt.[2] By the symbol $CaSO_4 + x\, MnSO_4$, is meant a solid solution of a trace of $MnSO_4$ in a matrix of $CaSO_4$.

[1] Van't Hoff; Ztschr. f. physik. Chemie, 5, p. 322, 1890.
[2] Wied. Ann., 56, p. 209, 1895.

Name of substance.	Colour of phosphorescence.
Pure $CaSO_4$.	Faint orange.
$CaSO_4 + x\,MnSO_4$.	Bright green.
Pure $SrSO_4$.	Does not phosphoresce.
$SrSO_4 + x\,MnSO_4$.	Bright red.
Pure $BaSO_4$.	Faint dark violet.
$BaSO_4 + x\,MnSO_4$.	Dark blue.
Pure $MgSO_4$.	Red.
$MgSO_4 + 1\,\%\,MnSO_4$.	Intense dark red.
Pure $ZnSO_4$.	White.
$ZnSO_4 + 1\,\%\,MnSO_4$.	Intense red.
Pure Na_2SO_4.	Bluish.
$Na_2SO_4 + .5\,\%\,MnSO_4$.	Intense brownish yellow.
Pure $CdSO_4$.	Yellow.
$CdSO_4 + 1\,\%\,MnSO_4$.	Intense yellow.
$CaFl_2$.	Faint blue.
$CaFl_2 + x\,MnH_2$.	Intense green.

Another illustration that the impact of cathode rays produces changes in the body on which the rays impinge, which last for a considerable time, is afforded by Crookes'[1] observation that glass which has been phosphorescing for a considerable time seems to get tired and to respond less readily to the action of the cathode rays. Thus, for example, in the experiment shown in Fig. 19, in which the shadow of a mica cross is thrown on the

---

[1] Crookes, Phil. Trans., pt. ii., 1879, 645.

walls of the tube, if after the experiment has been proceeding for some time the cross is shaken down, or a new cathode used whose line of fire does not cut the cross, the pattern of the cross will still be seen on the glass, but it will now be brighter than the adjacent parts, instead of darker. The portions outside the pattern of the cross have got tired by their long phosphorescence, and respond less vigorously to the stimulus than the portions inside

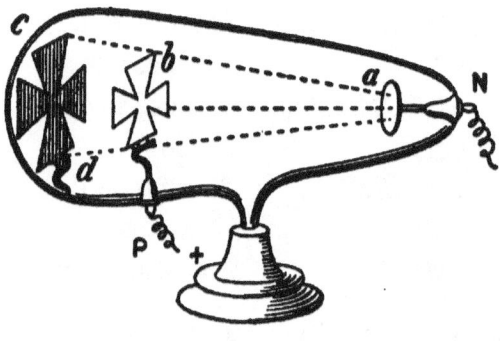

Fig. 19.

the original shadow which were previously shielded. Crookes found that this exhaustion of the glass could survive the fusing and reblowing of the bulb.

In many cases the phosphorescence on the glass is not of uniform intensity, but when carefully examined shows a granulated pattern full of detail. If the cathode is a flat disk of aluminium at the end of a long wire, the tube being cylindrical with the wire along the axis, then on the portion of the tube which phosphoresces under the rays coming from the straight wire, fine bands, alternately bright and dark, can often be observed, somewhat resembling the light and dark bands seen in straight diffraction fringes. These bright and dark bands, and the reticulated pattern caused by the cathode rays which come from the disk, are not due to a want of uniformity in the glass of the tube, for they can be moved about on the glass by means of a magnet. It seems possible that they may be due to irregularities on the cathode, perhaps produced by the discharge itself, as a cathode after long use has a roughened and irregular surface, — the long straight bands produced by a straight wire electrode being due to fine parallel scratches on the wire.

Thermal Effects produced by the Rays.

The rays heat bodies on which they fall; and by using a curved surface for the negative electrode, such as a portion of a hollow cylinder or of a spherical shell, this effect of the negative rays may be concentrated to such an extent that a platinum wire placed at the centre becomes red hot, and a piece of thin glass gets fused when the rays are directed on it.

Mechanical Effects produced by the Cathode Rays.

Mr. Crookes [1] has shown that when these rays impinge on vanes mounted like those in a radiometer the vanes are set in rotation. This mechanical action of the rays is well illustrated in the experiment represented in Fig. 20, where the axle of the vanes is mounted on glass rails; when the

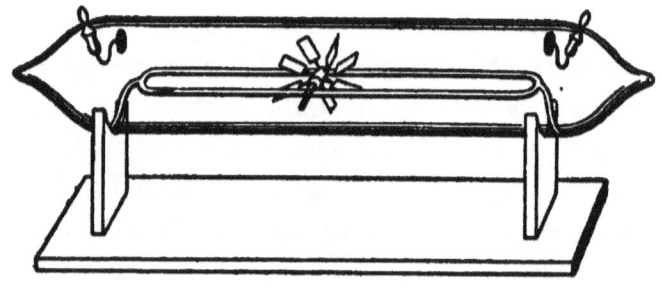

FIG. 20.

discharge passes through the tube, the cathode rays strike against the upper vanes, and the wheel travels from the negative to the positive end of the tube. In interpreting these experiments, we must remember that since the sides of the vanes exposed to the bombardment get hotter than the opposite sides, the vanes act, to some extent at

[1] Crookes, Phil. Trans., 1879, pt. i., p. 152.

any rate, like those of a radiometer, and thus the mechanical effects are complicated by thermal ones.

In another experiment due to Mr. Crookes the vanes are suspended as in Fig. 21, and can be screened from the negative rays by the screen *e*;

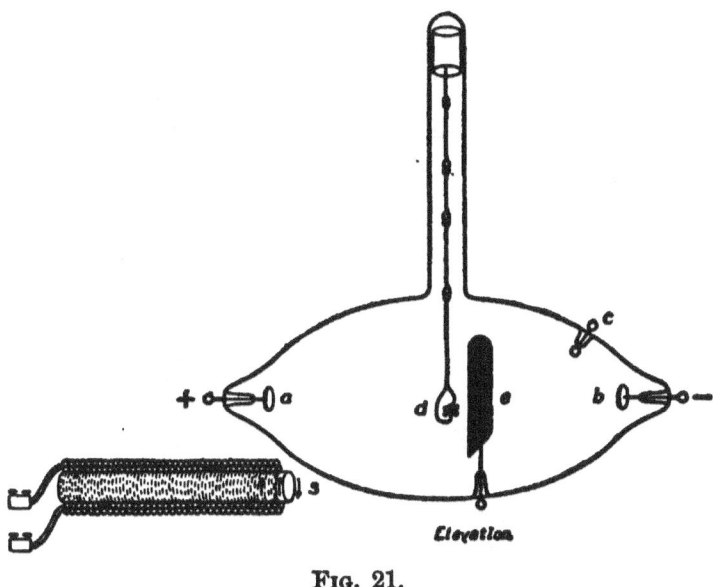

Fig. 21.

by tilting the tube the vanes can be brought wholly or partially out of the shadow of the screen. When the vanes are completely out of the shadow, they do not rotate, as the bombardment is symmetrical; when, however, they are half in and half out of the shadow, they rotate in the same direction as they would do if exposed to a bombardment from the negative electrode.

The mechanical effect produced by cathode rays is well shown by the following experiments due to

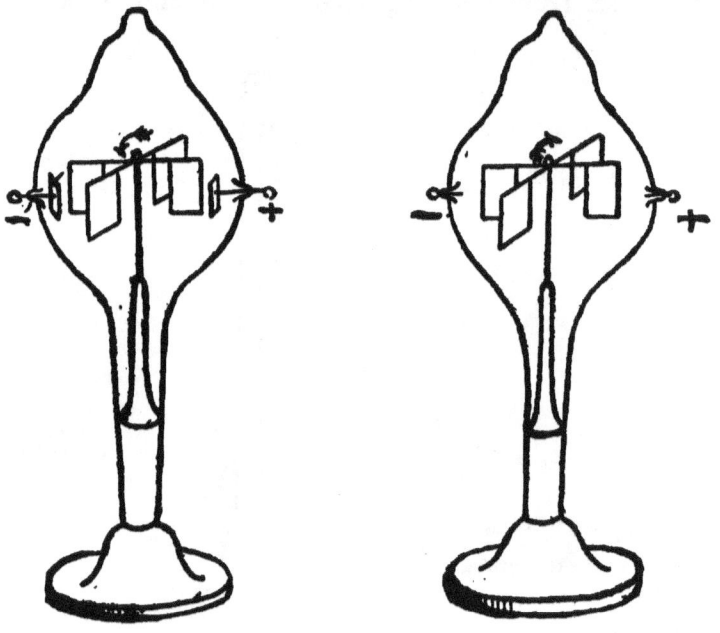

Fig. 22.

Puluj.[1] The apparatus is represented in Fig. 22. A small mica wheel is suspended on a needle-point by means of a small glass cap, and two cylindrical electrodes placed eccentrically are fused into

---

[1] Puluj, Radiant Electrode Matter, Physical Societies' Reprint of Memoirs, p. 275.

CATHODE RAYS

the glass vessel. Fig. 23 represents a section of the apparatus; the wheel revolves in the direction in which the cathode rays travel; that is, from the negative electrode. If the flat electrodes are replaced by wires, the mica wheel rotates in the opposite direction. Fig. 24 will show that this is in accordance with the previous experiment, as

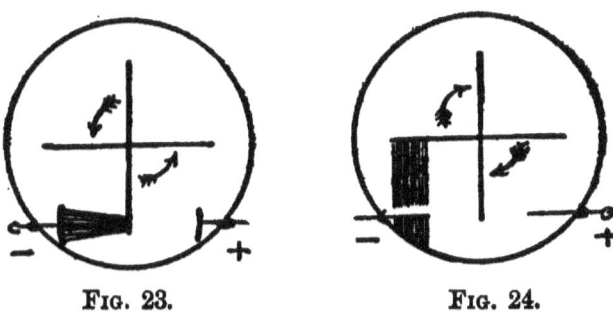

FIG. 23.   FIG. 24.

with the wire electrodes most of the cathode rays start from the sides, and not from the tip.

Both Crookes and Puluj have experimented using the wheel of a radiometer as the negative electrode, one side of the vanes being covered with mica. The interpretation of the results obtained is complicated, as there are several secondary causes at work to produce rotation besides the direct effect of the cathode rays. Thus, in consequence of the discharge, the sides of the vanes which are not covered with mica get hotter than the covered sides; this will tend to send the vanes round with the mica side in front. Again, if the rays strike the glass wall of the tubes, this will get unequally heated, and will, therefore, tend to set the vanes in rotation; that these secondary causes do produce considerable effect is shown by the observation of Puluj, that the rotation of the vanes continues for some time after the discharge has stopped. In addition, however, to these secondary effects, Crookes's experiments seem to show that when the dark space reaches the walls of the tube, a mechanical effect is produced of the same character as if the dark space were occupied by an elastic substance attached to the electrode.

Action of a Magnet upon the Cathode Rays.

A very interesting property of these rays is that they are deflected by a magnet so that the distribution of phosphorescence over the glass in the neighbourhood of the cathode, and the position of shadows cast by opaque bodies on the walls of the tube, are displaced when a magnet is brought near the tube.

The laws of the action of a magnet upon the cathode rays were first investigated by Plücker and subsequently by Hittorf and by Crookes. The result of these investigations can best be expressed by saying that they are in accordance with the view that these rays mark the path of particles charged with negative electricity, for such particles would in a magnetic field be acted upon by a force at right angles to their direction of motion and also to the magnetic force, and whose magnitude is proportional to the product of the magnetic force, the velocity of the particle, and the sine of the angle between the direction of these quantities. Under these circumstances the path of a particle in a uniform magnetic field would be a spiral wound on a cylinder whose axis was parallel to the direction of the magnetic force. In the particular case when the direction of projection is at right angles to the magnetic force, the path of the particle would be a circle in a plane at right angles to the magnetic force.

I have recently made a series of quantitative measurements of the effect of a magnet on these rays and have found the apparatus represented in Fig. 25 very useful for this purpose. The cathode was the disk, A, placed in a side tube which was fastened to the bell-jar by sealing-wax; the base of the bell-jar was a slab of plate-glass which was also fastened to the bell-jar with sealing-wax. The opening between the side tube and the bell-jar was closed by a tightly fitting solid brass cylinder with a slit cut in it. Care was taken to make this fit so tightly that the slit was the only opening between the side tube and the bell-jar. This cylinder was used as the anode and kept connected with the earth. The cathode rays start from the disk, go into the bell-jar through the slit in the cylindrical plug; in the bell-jar they pass in front of a ver-

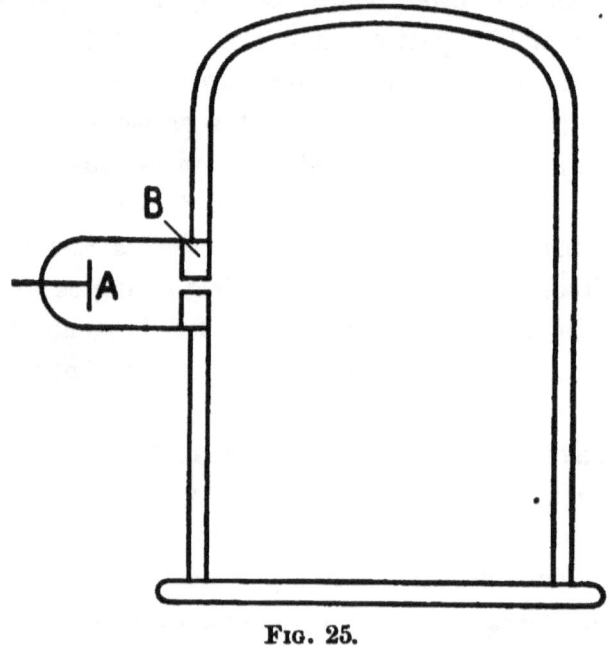

Fig. 25.

tical glass plate ruled by a diamond into squares whose sides were .5 cm. These squares render it easy to plot the path of the rays.

The tube with the bell-jar attached was placed between two large coils arranged as in Helmholtz's galvanometer, so that the distance between the planes of the coils was equal to the radius of the coils. This arrangement gives, as is well known, a very uniform magnetic field. The coils were arranged so that their planes were parallel to the path of the cathode rays; the magnetic force due to these coils would thus be at right angles to the path of these rays.

It was found that the cathode rays in their course through the gas produced sufficient luminosity to allow of their being photographed, provided the exposure lasted from 20 to 30 minutes; to photograph the squares on the plate a piece of magnesium wire was burnt.

The discharge was produced by means of an induction coil; the mean difference of potential between the cathode and the anode was measured by a Kelvin vertical voltmeter which was kept connected with these electrodes: the current through the coils used to produce the magnetic field was measured by an ammeter.

A specimen of the photographs taken with this apparatus is shown in Fig. 26.

There are some points which are very noticeable, whatever the pressure. In the first place, what, when there was no magnetic field, was a thin pencil of rays, spreads out in the magnetic field into a fan-shaped system of rays which may produce phosphorescence over two or three inches of the glass of the bell-jar, though when there is no magnetic force, the width of the phosphorescent patch is only about 3/16 of an inch wide.

It would thus appear that the original pencil of rays consists of a mixture of rays of different kinds,

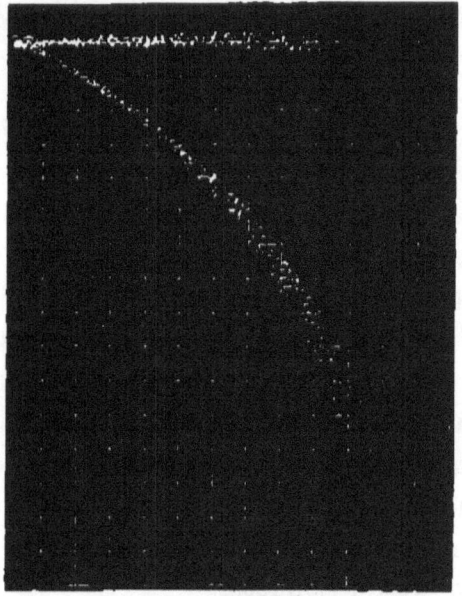

Fig. 26.

which suffer different deflections in the magnetic field.

Again, when the rays are spread out by the magnetic field, the luminosity seems to have a tendency to condense along certain lines, so that the luminosity in the gas had a streaky appearance, bright and dark streaks alternating. This effect was even more noticeable in the phosphorescence on the glass than in the luminosity in the gas, the phosphorescence in the glass showing a series of bright patches on a more faintly luminous background. This streaky appearance of the phosphorescence on the glass was first observed by Birkeland[1] and called by him the "magnetic spectrum," as the appearance resembles a spectrum with a series of bright bands in the green.

An examination of the photographs shows that the brightness of the phosphorescence produced on the glass at the end of the ray is not proportional to the brightness of the luminosity produced along the path of the ray through the gas. In some cases, indeed, when the phosphorescence on the glass was the brightest, the luminosity in the gas produced by the corresponding ray was so faint that it could not be detected on the photograph. Generally speaking, the rays which were the most deflected by a magnet were those which produced the greatest luminosity in the gas.

Paths of the Rays in Different Gases.

Many photographs of the paths of the rays in air, in hydrogen, in carbonic acid gas, in methyl iodide, were taken; and it was found that if the paths of the rays were compared, not at the same

[1] Birkeland, Comptes Rendus, 1896, p. 492.

pressure, but at such pressures as gave the same mean potential difference between the electrodes, then the paths of the rays were identical in all gases. This identity was not confined to the rays which suffered the greatest deflection, but all the details, such as the distribution of the bright and dark streaks, the positions of the brightest spots on the glass, were the same in all cases, so much so, indeed, that in some cases the photograph of the rays in one gas could not be distinguished from that of the rays in another.

We may compare this result with that found by Lenard for the Lenard rays; the deflection of these rays produced by a magnet is independent of the nature and pressure of the gas through which they are passing.

When the pressure of the gas in the discharge tube in which the cathode rays are produced is diminished, the average potential difference between the anode and cathode increases, and rays which are less deflected than those reproduced at higher pressures appear. The appearance presented when the pressure diminishes suggests rather the introduction of rays which are less deflected and a diminution in the intensity of the more easily deflected rays, than a mere diminution in the deflection of the rays present at the higher pressure, as traces of the more easily deflected rays remain after the reduction of the pressure; that is, the lower boundary of the phosphorescent patch on the glass of the bell-jar seems to fade gradually away rather than to rise when the pressure is diminished.

If an air break is placed in series with the discharge tube, the place on the glass where the phosphorescence is brightest approaches the undeflected position of the phosphorescent spot more and more closely as the length of the air break increases; traces, however, remain of the more deflected rays, which are predominant when there is no air break, and there is no shifting up of the extremity of the phosphorescent patch corresponding to the greatest deflection. This edge, however, becomes much fainter after the insertion of the air break.

The curvature of the cathode rays at any point seems to be a function of the magnetic force at that point. Thus, when the magnetic field through which the rays pass is that due to the two coils, and a horse-shoe magnet is placed so as to counterbalance at a certain point in the bell-jar the magnetic force due to the two coils, the path of the cathode rays has a point of inflection at the point where the magnetic force vanishes.

All the photographs of the paths of cathode rays in a magnetic field contain rays stretching a long way across the bell-jar, and which have not suffered any deflection by the magnet. These rays are present through a wide range of pressures, though they get faint when the pressure is very low. These rays do not seem to possess that power of producing phosphorescence on solids on which they fall, which is so characteristic of the cathode rays. I was never able to feel sure that any phosphorescence at all was produced by these rays when they fell on the glass of the bell; and if any such existed it must certainly have been exceedingly faint compared with that produced by the rays which were deflected by a magnet. To test this point further, a screen over which a thin membrane was stretched was placed in the bell-jar near to the slit (the membrane was an animal one, obtained, I believe, from the stomach of the calf, and was thinner than any tissue paper I have ever seen). The rays all fell on the membrane, and the arrangement was photographed, a long exposure being given. No trace of any rays, whether deflected or not, could be detected on the far side of the screen; but the parts of the membrane struck by the deflected rays phosphoresced brightly with a greenish phosphorescence, while the places struck by the undeflected rays did not show any phosphorescence at all.

Goldstein [1] found that with a perforated cathode certain rays occurred near the cathode which were

[1] Goldstein, Berliner Sitzungsberichte, 39, p. 691, 1886.

not deflected by a magnet; from their method of production he called them "Kanalstrahlen;" the undeflected rays seen in Fig. 26 may be very finely developed "Kanalstrahlen." No property other than that of being accompanied by luminosity in the gas is, as far as is known at present, possessed by these rays; and it is possible that they are jets of phosphorescent gas forced through the slit by a kind of explosion at the cathode.

The photographs very frequently show a short strongly curved streak of luminosity situated above the rays that come through the slit. This, I think, is an example of the phenomena discovered by Goldstein,[1] that when the electric discharge passes through a body pierced with holes, each of the holes acts as a secondary cathode and emits cathode rays. The same thing was shown also by Goldstein to occur when there is any considerable constriction in the cross section of the discharge tube. In the experiment from which the photograph is taken, the slit in the plug separating the side tube from the bell-jar seems to act like the perforation or constriction in Goldstein's experiments.

When the pressure is reduced as low as possible compatible with there being any discharge at all, more cathode rays seem to come off from the plug acting as a secondary cathode than pass through

[1] Goldstein, Wied. Ann., 11, p. 832, 1880.

the slit itself. This causes a very curious reversal of the phosphorescent pattern on the glass at higher pressures: when the majority of the rays come through the slit, the phosphorescent patch is a rectangular strip corresponding to the slit; at the lowest pressures, however, when the rays seem to come chiefly from the plug itself, the part corresponding to the slit is dark.

In some of these experiments there appeared at very low pressures a phosphorescent pattern on the glass somewhat elliptical in outline and very bright. Its position was such that the cathode rays which produced it could not have come through the slit; they appeared to come from the edges of the slit, on the side of the plug furthest from the cathode. They immediately disappeared if the plug was connected with the earth. The most noteworthy feature of this phosphorescence was that it lasted for some time, two or three minutes, after the coil was stopped. It was deflected by a magnet like the phosphorescence produced by the ordinary cathode rays. The fact that the cathode rays carry charges of negative electricity will, I think, explain this effect. The cathode rays with their negative charges strike against the sides of the tubes and against the plug, and the negative electricity accumulates until the density of the electrification gets so great that the negative electricity discharges and forms cathode rays on its own account. As it takes some time for the rays to discharge the negative electrification, they continue to flow for some time after the coil has stopped.

The accumulation of negative electricity carried by the cathode rays would also account for a perforation in a solid obstacle or a constriction in the cross section of the tube acting as a cathode.

Shape of the Path of the Rays.

The photographs show that the path of the rays producing luminosity in the gas when acted on by the uniform magnetic field is approximately circular. If the pressure is somewhat high, then after the rays have travelled some distance the curvature of their paths seems to increase. This is what would happen if the cathode rays consisted of electrified particles projected from the cathode through the bell-jar, as in consequence of collisions with other molecules the velocity of the particles would gradually diminish and the curvature of their paths consequently increase.

Electric Charge carried by the Cathode Rays.

The following experiment, which was made by Perrin,[1] shows that a current of negative electricity

[1] Perrin, Comptes Rendus, 121, p. 1130, 1895.

flows along the path of the cathode rays. The apparatus is represented in Fig. 27. The cathode is a disk; the anode which is connected with the earth is a metal cylinder pierced with a hole on the end of the cylinder turned to the cathode. Inside this cylinder and insulated from it is a second coaxial cylinder, also with a hole through the end turned to the cathode; the line joining this hole to the one in the outer cylinder is at right angles to the

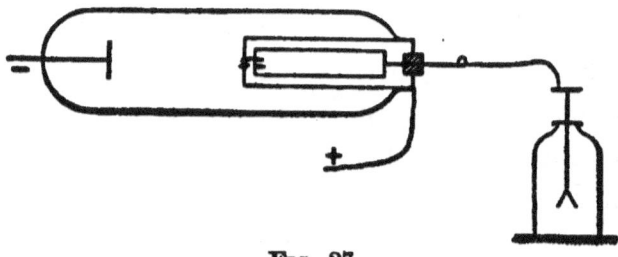

Fig. 27.

cathode, so that the cathode rays can penetrate into the inside of the inner cylinder. The inner cylinder is connected with a gold-leaf electroscope. When the cathode rays penetrate into the inner cylinder, the gold leaves of the electroscope diverge with a negative charge. If, however, the cathode rays are deflected by a magnet so that they no longer pass into the inner cylinder, the gold leaves of the electroscope do not diverge, showing that in this case there is no electrification inside the cylinder. Thus the cylinder receives a negative charge when the cathode rays enter it, while it gets no charge when they do not. Hence it would seem that the cathode rays are accompanied by a charge of negative electricity. Perrin found that the negative charge carried by these rays was very considerable. Thus in one of his experiments the charge sent into the inner cylinder for each interruption of the coil was sufficient to raise a capacity of six hundred electrostatic units to a potential of three hundred volts.

Perrin's experiment, though it shows that a negative charge follows the course of the cathode rays when these are undeflected, does not show that the path of the negative charge coincides with the path of the cathode rays when these are deflected by a magnet. It is open to the objection that the charge conveyed to the cylinder might be due to electrified particles shot off at right angles to the cathode, and so coinciding in path with the cathode rays when these are not deflected by a magnet: as these charged particles would be deflected by a magnetic field, the cylinder would not be charged when a magnet was brought up to the discharge tube; and yet these charged particles might only be accidental accompaniments of the cathode rays, and not necessarily essentially connected with them, following them in every twist or turn they might make: just as a cannon-ball may accompany the discharge of a cannon, though it is not essential to the flash. To meet these objections, I modified Perrin's experiment, using the arrangement figured in the following diagram.[1]

The rays start from the cathode, *A*, and pass through a slit in a solid brass rod, *B*, fitting tightly

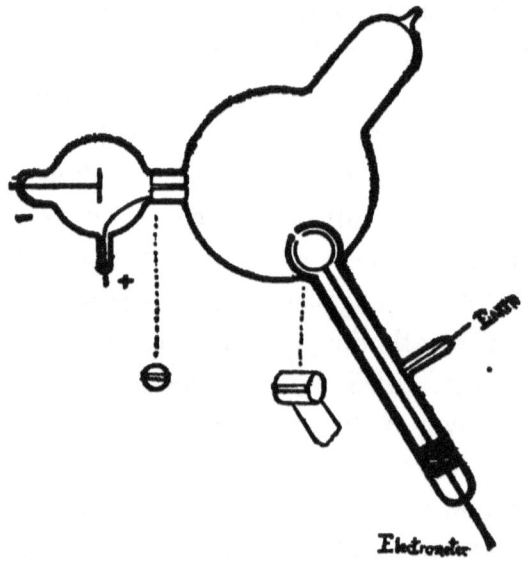

FIG. 28.

into the neck of the tube. This plug is connected with earth and used as the anode; the rays after passing through the slit travel through the vessel *C*. The arrangement used to measure the charge consists of two coaxial metal cylinders, *D* and *E*,

[1] J. J. Thomson, Proc. Camb. Phil. Soc., ix., 1897.

insulated from each other and each having a slit cut in it to enable the rays to pass into the inside of the inner cylinder. The inner cylinder is connected with an electrometer, the outer cylinder with the earth. The two cylinders are placed on the far side of the vessel, but out of the direct line of fire of the rays. With this arrangement there was, when the induction coil was turned on, a small but only small negative charge in the inner cylinder: sometimes, indeed, at extremely low pressures the charge in the inner cylinder was positive; this was, I think, due to the outer cylinder acting as a cathode and giving a positive charge to the inner one. The cathode rays were then gradually deflected by a magnet, and when the phosphorescent patch (which was drawn out by the magnet to a considerable width) fell upon the slit in the outer cylinder, the inner cylinder received a large negative charge. The increase in the negative charge coincided very sharply with the appearance of the phosphorescent patch on the slit; when the phosphorescent patch had been so much deflected by a magnet that it fell below the slit, the negative charge in the inner cylinder again disappeared. This experiment shows that the negative electrification follows exactly the same course as the rays which produce the phosphorescence on the glass.

The experiment also shows that, in addition to

the negative electrification which follows the course of the cathode rays, the passage of these rays through the gas puts it into a state in which it is a conductor of electricity. This is shown by the fact that if we keep the induction coil running, and direct a stream of cathode rays steadily into the inner cylinder, the negative deflection of the electrometer reaches a certain value, beyond which it does not increase, however long the rays may be kept running into the cylinder. Again, if the inner cylinder is charged positively before the rays fall upon it, as soon as the rays are turned on, the cylinder will rapidly lose its positive charge and acquire a negative one. If the cylinder has initially a small negative charge, this, on the other hand, will increase when the cathode rays play upon the cylinder. If, however, the initial negative charge is very large, the cylinder will rapidly lose a portion of this when the rays play upon it. Whatever may be the charge on the cylinder, the leak stops when the rays are cut off. If the cathode rays merely carried a charge of negative electricity, the cylinder would not go on losing a negative charge as long as the rays were on, and would retain whatever charge might be left as soon as the rays were cut off.

These experiments show that when the cathode rays pass through a gas they make it for the time being a conductor of electricity.

This result will enable us to explain why, though, as we have seen, the rays carry charges of negative electricity, yet they are not deflected directly by an electrostatic field. The absence of an effect of electrostatic force on the path of the rays is shown by the following experiment which is identical in principle with one made by Hertz.[1] To the glass plate used in the experiments to determine the paths of the rays in a magnetic field, Fig. 25, p. 151, two parallel pieces of metal were attached in such positions that the cathode rays, after passing through the slit in the plug, travelled between those strips. Wires fused to these strips passed through the bell-jar, and were connected with the terminals of a battery of small storage cells, so that the two strips can readily be charged up to as great a difference of potential as they will bear without sparking. The direction of the electric force is at right angles to the strips, and thus at right angles to the paths of the rays; the electric field would thus tend to move the negative charges up or down, and so produce a displacement of the patch of phosphorescent light on the glass. On trying the experiment no appreciable difference could be detected between the position of the phosphorescent patch when the metal strips were at their maximum potential difference, and when

[1] Hertz, Wied. Ann., 19, p. 782, 1883.

they were connected together so as to be at the same potential.[1] It must be noticed, however, that when the rays are passing between the plates a discharge readily passes between them, so that the maximum potential difference which can be established between the metal strips is only a small fraction of that which could be established if the gas were in its normal state. Charged bodies may be brought up to the rays without producing any deflection; and if, as sometimes happens, the phosphorescent patch is deflected when the slit through which the rays pass is disconnected from earth, the effect seems to be due to the production of a second stream of cathodic rays and the deflection of each in accordance with the result described on page 168. Thus the negative charges in their course through the body of the gas do not seem to be deflected by an electric field. This would be what we should expect, from the fact that the cathode rays make the gas through which they pass a conductor, so that the negative electricity is travelling through a conducting medium, and we should no more affect the path of the rays by an external electric force than we should the path of a current in an electrolyte by holding an electrified body in its neighbourhood. It may, however, be urged

[1] On repeating this experiment recently I find that at low pressures there is a deflection of the rays.

that when we have the metal strips actually immersed in the gas and connected with a battery of given electro-motive force, we produce an electric intensity equal to the potential difference between the plates divided by the distance between them, whether the gas is a conductor or not. We must remember, however, that the rays do not fill up the whole of the space between the strips, and that the gas just around the rays is probably an infinitely better conductor than that outside, so that the potential gradient in the region close to the electric charges is probably only an infinitesimal fraction of the value deduced on the assumption that the medium between them is of uniform conductivity.

Repulsion of Cathodic Streams.

Goldstein[1] discovered that if in a tube there are two cathodes connected together, the cathodic rays from one cathode are deflected when they pass near the other cathode.

This repulsion is illustrated by the following experiment, which is also due to Goldstein: a flat piece of metal with two holes in it is placed in the discharge tube between two parallel wires which can be used as cathodes; the plane of the metal being at right angles to the plane containing the

[1] Goldstein, Eine Neue Form der Elektrische Abstossung.

two cathodes, this plane passes through the middle line of the two holes in the plate. Let us suppose that the metal plate is parallel to the plane of the paper, and that one of the wires, b, is in front of this plane and the other wire, a, behind it. When a is cathode, the rays from it throw two bright phosphorescent patches on the walls of the tube; these patches are crossed by a dark line, which is the shadow of b. Fig. 29 a. When, however, b is connected with a so that both are cathodes, the two halves of these bright patches are repelled from each other and present the appearance shown in Fig. 29 b. Here again the effect can be explained by supposing that the rays from a which pass near b are repelled from it when it is a cathode.

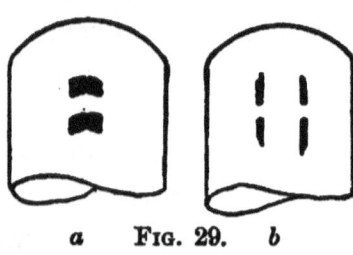

a Fig. 29. b

The deflection of the rays themselves in their course through the glass, and not merely of the phosphorescence produced by them on the glass, was shown by Goldstein in the following way: one of the cathodes was a hollow metal cylinder, Fig. 30, from the inside of which a pencil of cathode rays proceed. The course of these rays can be traced by the luminosity they produce in the gas. The second cathode was a wire, b, at right angles to the cathode rays proceeding from a. When b is disconnected from a, the path of the cathode rays is straight; but when b is connected with a, the cathode rays from a bend sharply round when they approach b, as in Fig. 30. The bending is so sharp that the cathodic rays from a look almost as if they had a nick in them.

Goldstein found that the magnitude of the deflection did not depend on the materials of which the electrodes were made, nor on the nature of the gas through which the rays passed. The deflection was stopped, if the deflecting cathode was surrounded by a screen of some solid substance.

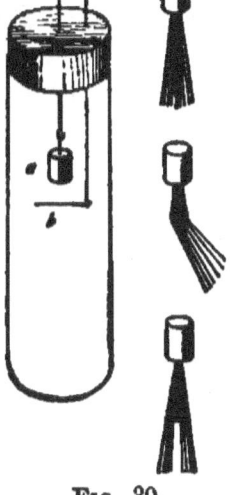

Fig. 30.

An experiment made by Crookes[1] and represented in Fig. 31 seems to be an example of this effect. a and b are metal disks, either or both of which may be made into cathode; a diaphragm with two holes, d and e, cut in it, is placed in front of the disks, and the path of the rays is traced by the phosphorescence they excite in a chalked plate inclined at a small angle to their path. When a

[1] Crookes, Phil. Trans., 1879, pt. ii., p. 652.

is the cathode and *b* is idle, the rays travel along the path *df*; and when *b* is the cathode and *a* is idle, they travel along the path *ef*. When, however, *a* and *b* are cathodes simultaneously, the paths of the rays are *dg* and *eh* respectively, the two rays having thus repelled each other. Crookes attributed the divergence of the rays to the repulsion between the negative charges of electricity travelling along them. The absence of deflection when an electrified body is brought into the neighbourhood of the

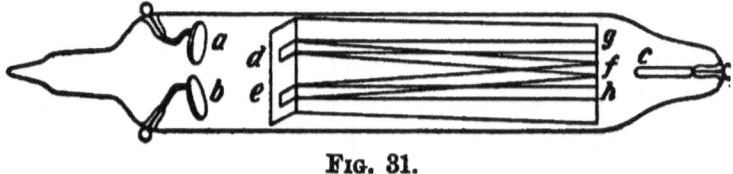

Fig. 31.

tube shows that the gas in the body of the tube is in a conducting state, and so would screen off from one set of rays the electric force due to the other set. Indeed, E. Wiedemann and Ebert[1] have shown directly by a modification of Crookes's experiment that the divergence of the rays is not due to the repulsion between the electric charges travelling along them. In this modification of the experiment, the holes *d* and *e* are provided with shutters. Wiedemann and Ebert found that when *a* and *b* were simultaneously cathodes, *eh* was the path of the rays

[1] Wiedemann and Ebert, Wied. Ann., 46, p. 158, 1891.

through e, whether the window at d was open or shut, though of course when it was shut there were no cathode rays travelling along dg. The deflection of the rays seems to be an example of Goldstein's discovery, that a cathode repels the rays proceeding from another cathode.

The repulsions discovered by Goldstein seem to me to admit of explanation on the view that they are due to the electrostatic repulsion of the negative electricity travelling along the cathode rays by the negatively electrified cathode, this repulsion being exerted only in the neighbourhood of the cathode, since the gas in the other parts of the tube is a conductor and incapable, therefore, of transmitting electrostatic repulsions. On this view the gas in the neighbourhood of the cathode is in a different state from that in the rest of the tube: in the neighbourhood of the cathode the gas is an insulator; in the rest of the tube it is a conductor.

The difference between the properties of the gas in the neighbourhood of the cathode and the rest of the gas in the tube is shown by the following experiment. In Fig. 32, A represents a large bulb, and B a small one connected by the tube C. Two wires, a and b, are fused into the bulbs, A and B respectively, and can be used as cathodes; the anode is the wire c. The bulbs A and B are wound with coils of wire placed in series and connected with the outsides of two Leyden jars. The insides of these jars are connected with the terminals of a Wimshurst machine; when sparks pass between these terminals, the jars discharge, and electric currents pass through the coils connecting their outer coatings. These currents alternate very rapidly and produce, therefore, electro-motive forces in the regions adjacent to them. If the pressure of the gas in the bulbs A and B is very low, these electro-

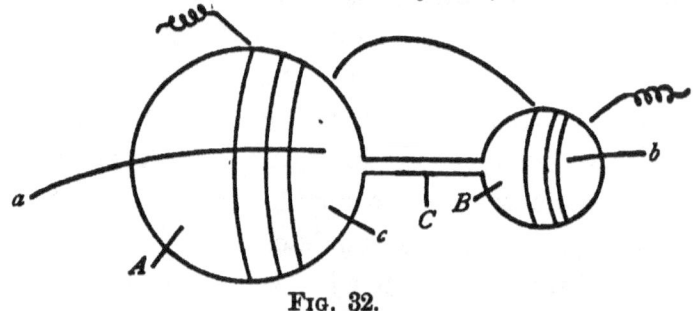

FIG. 32.

motive forces cause a discharge to pass through the gas. The discharge produced in this way is the ring discharge. (See Recent Researches in Electricity and Magnetism, p. 92.)

The bulbs are pumped until the pressure is such that when a and b are cathodes the dark space completely fills the small bulb, B, while it does not reach to the walls of the larger bulb, A. The induction coil is stopped, and the ring discharge is sent through the bulbs A and B, the length of spark given by the Wimshurst being adjusted so that the brightness of the ring discharge is such that a change in its intensity can readily be detected. The coil is now turned on, and it is found that the ring discharge stops completely in the small bulb, which is wholly occupied by the dark space; while in the larger bulb the ring discharge, which passes outside the dark space, increases very much in brightness. It thus appears that the conductivity of the gas inside the dark space is diminished, and that outside increased, by the passage of the discharge. The conductivity of a gas in the neighbourhood of a cathode may be small enough to admit of electrostatic repulsion, and thus give rise to the repulsion between two cathodic streams which we have been describing.

Wiedemann and Ebert[1] have shown that the dark space not only offers a great resistance to the passage through it of cathodic rays, but that it is distorted when such rays fall upon it.

The repulsion of adjacent cathodic streams seems to have a great deal to do with the product of phosphorescent patterns of great interest, which Goldstein[2] obtained by using cathodes of different shapes. In many of these experiments he used cathodes forming part of the surface of a

[1] Wiedemann and Ebert, Sitzungsberichten der physikal med. Societät zu Erlangen, Dec. 1891, p. 31.
[2] Goldstein, Wied. Ann., 15, p. 254, 1882.

sphere 40 mm. These were placed in a vessel whose diameter was twice the radius of curvature of the cathode. Under these circumstances, if the cathode rays came off accurately at right angles to the surface of the cathode, the phosphorescent pattern would be the same shape and size as the

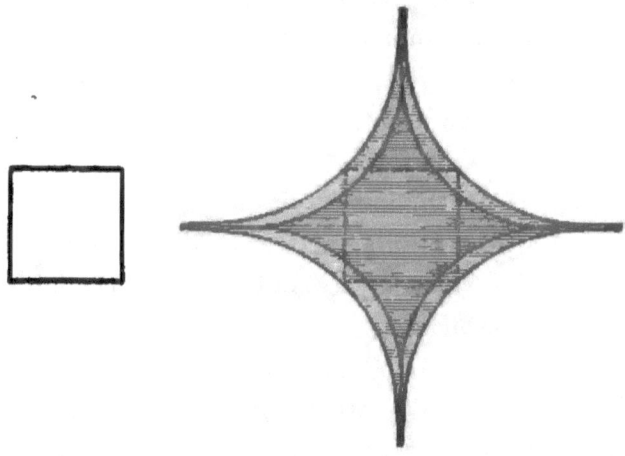

Fig. 33.

cathode, but turned upside down. At comparatively high pressures this was found to be the case; but when the pressure was exceedingly low, this was not even an approximation to the truth.

Thus, when the boundary of the curved cathode was a square, Goldstein obtained at very low exhaustions the phosphorescent pattern shown in Fig. 33, where the dotted lines indicate what the outline of the pattern would be if all the cathode rays came off at right angles to the cathode.

When the boundary of the curved cathode was a triangle, the pattern was that shown in Fig. 34. By covering up a portion of the triangular cathode with a screen, Goldstein showed that the top of the phosphorescent pattern was formed by rays coming from the top of the cathode; so that at

Fig. 34.

this low exhaustion the cathode rays have not crossed.

When the cathode was a curved rectangular cross, the pattern formed as the pressure was gradually diminished went through the series of changes shown in Fig. 35.

With a cross with an odd number of arms, the phosphorescent pattern was that given in Fig. 36. The pressure in this case was not very low, and we see from the shape of the figures that the rays must have crossed.

When the cathode was a plane cross, the phosphorescent pattern was that shown in Fig. 37.

The shapes of these patterns at high exhaustions, especially in the case of the plane cross,

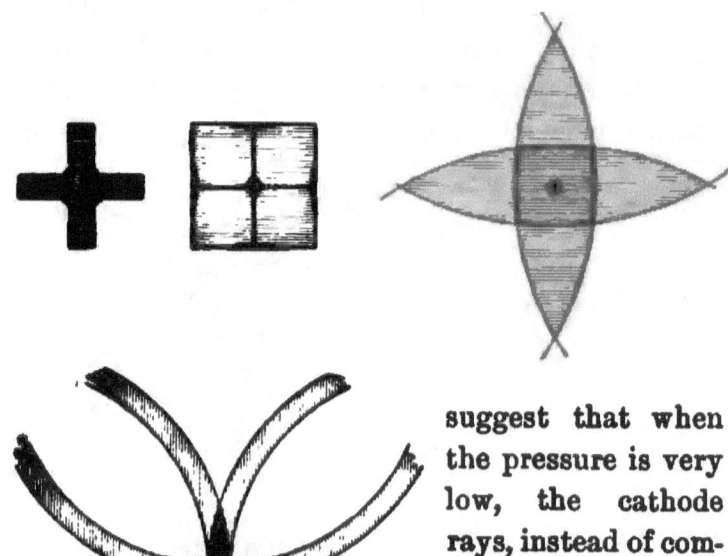

suggest that when the pressure is very low, the cathode rays, instead of coming off uniformly from the whole of the cathode, come off mainly from places where the surface density of the electricity is greatest; that is, from the edges of the cross, these edges forming, as it were, the effective cathodes; and that these cathodes exert on the

Fig. 35.

rays coming from the neighbouring cathodes the repulsive action described above.

Diffuse Reflection of Cathode Rays.

When the cathode rays fall upon a surface, whether that of an insulator or of a conductor, the substance struck generally becomes itself a cathode, and emits cathode rays mainly normal to its surface. In addition to this, and especially if the cathode rays strike the surface obliquely, cathode rays

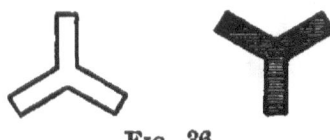

Fig. 36.

Fig. 37.

start from the surface in all directions. The phenomenon is regarded as analogous to the diffuse reflection of light from such a surface as gypsum, and the cathode rays are said to be diffusely reflected from the surface. The phenomenon can be very simply shown by the following experiment, which is due to Goldstein.[1] The tube is shaped like that shown in Fig. 38: k is the cathode, the rays from which fall upon

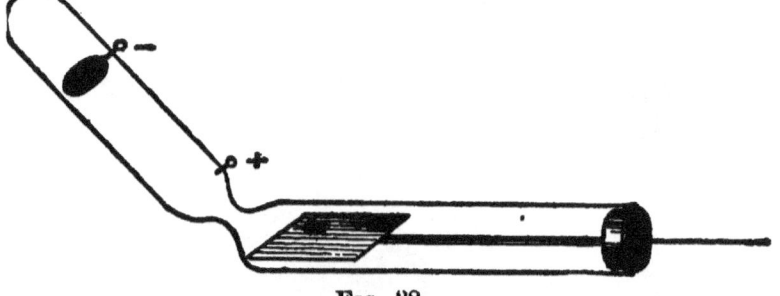

Fig. 38.

the plate f, which can be rotated by turning the rod which passes through the stopper at A. The walls of the tube AC become phosphorescent from the cathode rays diffusely reflected from f. This occurs even when the plate is made of some substance which does not of itself phosphoresce under the action of cathode rays.

It has not, I think, yet been proved that the diffusely reflected rays have the same properties

[1] Goldstein, Wied. Ann., 15, p. 254, 1882.

as the incident ones. It would be interesting to know if they suffered the same deflection when traversing a magnetic field.

Transmission of Cathode Rays.

It was stated on page 137 that a solid placed between the cathode and the walls of the tube cast a black shadow. Crookes and Goldstein have shown that bodies as thin as very thin glass or a film of collodion cast extremely black shadows. Hertz[1] found, however, that behind a piece of gold leaf or aluminium foil there was appreciable phosphorescence on the glass, and that the phosphorescence behind the gold leaf moved when a magnet was brought near. Hertz found that when two layers of gold leaf are placed one behind the other, the diminution in the brightness of the phosphorescence on the glass behind them is much greater than would be anticipated from the effect produced by one film of gold leaf. He ascribed this effect to the reflection of the phosphorescent light from the gold leaf. Thus suppose AB is the gold leaf, and that this reduces the brightness of the phosphorescence on the glass behind it to 1/3. An eye placed at E would only see a diminution to 2/3, if the gold leaf were a perfect reflector, for in addition to the light coming directly from the glass

[1] Hertz, Wied. Ann., 45, p. 28, 1892.

at P, it would also receive that reflected from the gold leaf. If, however, we had two layers of gold leaf, then the brightness of the phosphorescence on the glass would be reduced to 1/9, and even after reflection it would only amount to 2/9.

From the transmission of cathode rays through thin films of metal, Hertz concluded that they were waves in the ether, and that these passed through the thin films like waves of light through a slightly transparent medium.

Lenard's Experiments.

A series of most important experiments on the passage, or apparent passage, of cathodic rays through thin films was made by Lenard,[1] who used a tube like that shown in Fig. 39. In this tube, K, the cathode is an aluminium disk, 12 mm. in diameter, fastened to a stiff wire which is surrounded by a glass tube. The anode A was a brass tube surrounding part of the cathode.

The end of the tube opposite the cathode was closed by a strong metal cap fastened to the tube by marine glue; a hole 1.7 mm. in diameter was bored in the middle of the cap, and this hole was covered with a piece of thin sheet aluminium whose thickness was about .00265 mm. The window was in metallic contact with the cap, and this and the an-

[1] Lenard, Wied. Ann., p. 51, 225, 1894.

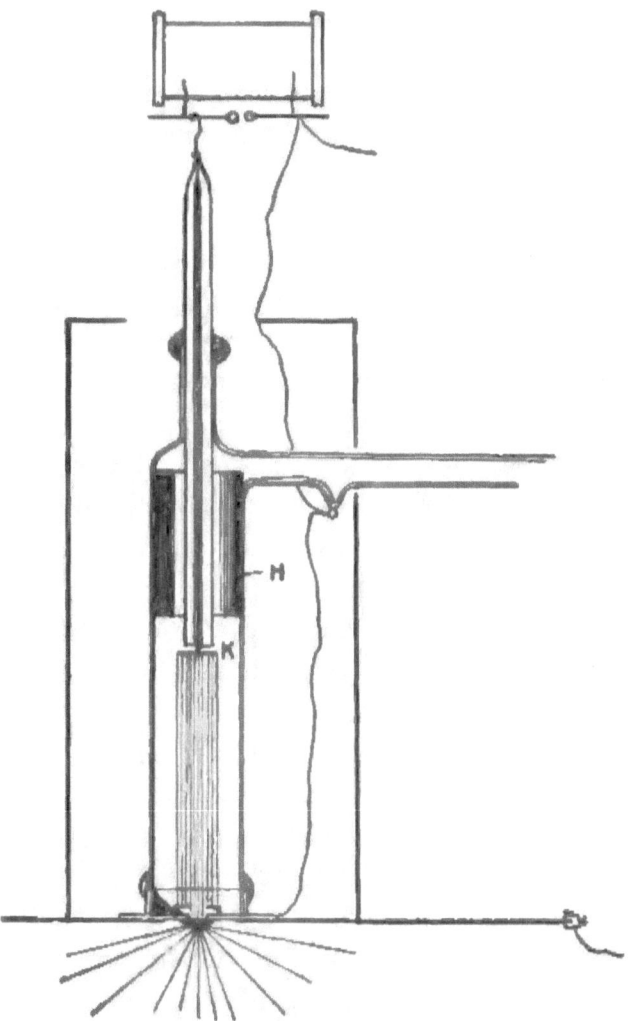

Fig. 39.

ode were connected with the earth. The tube was then exhausted until the pressure was so low that the cathode rays reached the window. Then in a dark room diffuse light is seen to spread from the window into the air outside the tube; this light is brightest close to the window, is without any definite boundary, and ceases to be visible at a distance of about 5 cm. from the window. At each discharge of the coil this light spreads like a brush discharge from the window.

Phosphorescent bodies placed in the neighbourhood of the window phosphoresce, and the brightness of their phosphorescence seems to depend only on the distance of the body from the window, and not on the direction of the line joining it to the window.

The most sensitive phosphorescent substance was found to be tissue paper soaked in a solution of pentadekylparatoleketone.

This phosphorescence outside is produced when the cathode rays from the inside of the tube strike against the thin aluminium window; rays which, like cathode rays, have the property of producing phosphorescence in bodies against which they strike start from the far side of the window. We shall find that the properties of the rays outside the tube resemble in all other respects cathode rays; as, however, it will be convenient to distinguish the rays outside the tube from those inside, we shall call the former Lenard rays.

The Lenard rays spread out very diffusely, so that the shadow cast by an opaque body on the bright phosphorescent patch seen when the rays pass through a hole in a solid is ill defined and many times larger than the geometrical shadow.

Lenard found that these rays produce photographic effects. They affect sensitized paper and photographic plates. Covering a photographic plate partly with a sheet of aluminium and partly with a thin piece of quartz, Lenard obtained the photograph shown in Fig. 40. From this photograph it will be seen that the transparent quartz has stopped the action, while the optically opaque aluminium has allowed it to pass through. When this photograph was taken, Röntgen rays had not been discovered; it is possible that part of this effect may have been due to these rays.

FIG. 40.

Lenard found that a positively or negatively electrified body lost its charge when the rays diverging from the window fell upon it. The interpretation of this result is rendered difficult from the fact that Röntgen rays produce the same effect; as, however, the cathode rays inside the tube render the gas through which they pass a conductor of electricity, we should expect Lenard rays which are cathode rays outside the tube to possess the same property.

If the aluminium window opposite the cathode, instead of opening into the air, opens, as in Fig. 41, on to another tube which can be exhausted, we can by means of it study the Lenard rays under various conditions.

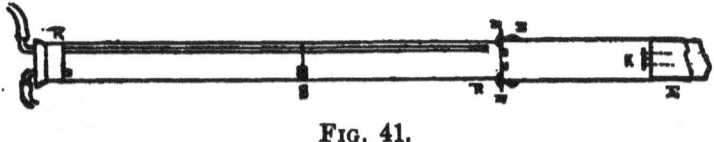

Fig. 41.

The lower the pressure of the gas in the tube (*S*), the further the rays travel, and the less diffuse they are. By filling the tube with different gases Lenard found that the greater the pressure of the gas, the greater the absorption of these rays. Thus the rays travel far further in hydrogen than in any other gas at the same pressure. Lenard found that if he adjusted the pressure so that the density of the gas in the discharge tube was constant, — if, for example, the pressure when the tube was filled with oxygen was 1/16 of the pressure when it was filled with hydrogen, — the absorption was constant, whatever the nature of the gas.

If the pressure in the tube B containing the cathode altered, the distance traversed by the Lenard rays in the tube A before their effects ceased to be visible altered, too. There are thus Lenard rays of different kinds. The lower the pressure in the cathode tube, the further the Lenard rays travel in the tube A. Thus, when the pressure in the cathode tube is high, Lenard rays which are easily absorbed are produced, while those produced when the pressure in the discharge tube is low can travel long distances. At very low pressures we know that a discharge refuses to pass from one terminal to another in a vacuum tube; if, however, the gas in the tube B is at such a low pressure as to stop the discharge; yet Lenard rays can pass through with great facility if the pressure in the tube A is high enough to allow the discharge to pass.

In a later paper Lenard[1] investigated the absorptive power for Lenard rays of solids as well as gases; the results are given in the table on the following page.

By the coefficient of absorption is meant a quantity k such that e^{-kh} represents the ratio of the intensity of the rays after traversing a thickness h of the substance to their initial intensity.

[1] Lenard, Wied. Ann., 56, p. 255, 1895.

Substance.	Coefficient absorption.	Density.	Absorption density.
Hydrogen at 3 mm. pressure	.00149	.000000368	4040
Air at .78 mm. pressure	.00416	.00000125	3330
Hydrogen at 760 mm. pressure	.476	.0000849	5610
Air at 760 mm. pressure	3.42	.00123	2780
SO_2 at 760 mm. pressure	8.51	.00271	3110
Collodion	3310	1.1	3010
Paper	2690	1.30	2070
Glass	7810	2.47	3160
Aluminium	7150	2.70	2650
Mica	7250	2.80	2590
Dutch metal	23800	8.90	2670
Silver	32200	10.5	3070
Gold	55600	19.3	2880

We see from this table that though both the coefficient of absorption and the density vary enormously, yet the quotient of the coefficient by the density only varies between the limits 2070 and 5610. This seems to point to the very remarkable result that the absorptive power of bodies for the Lenard rays depends only upon the density of the body, and is independent of its chemical composition and physical state.

Magnetic Deflection of the Rays.

Lenard[1] has studied the deflection by a magnet of the rays in the tube A. He finds that if the nature of the gas in B and its pressure is kept constant, then the curvature of the Lenard rays due

[1] Lenard, Wied. Ann., 52, p. 23, 1894.

to a given magnetic field is independent of the pressure and nature of the gas in A.

This curvature depends greatly, however, on the kind of gas in B and its pressure, a diminution of the pressure of the gas in B causing the magnetic deflection of the Lenard rays in A to diminish.

Thus the smaller the magnetic deflectibility of the cathode rays in B, the smaller the magnetic deflectibility of the rays they produce.

Theories of the Nature of the Cathode Rays.

Two very widely divergent views of the nature of the cathode rays. The one view, which was first advanced by Goldstein, and which has since been advocated by E. Wiedemann, by Hertz and by Lenard, is that the cathode rays are waves in the ether. The majority of physicists who support the ethereal theory of the cathode rays seem to regard these waves as transversal; a theory, however, has been advanced by Jaumann according to which these waves are longitudinal ones. We shall for brevity call the theories which regard the cathode rays as actions taking place in the ether as the "ether theories."

The other theory, which was first advocated by Crookes, regards the cathode rays as marking the course of a stream of negatively electrified particles, these particles, unless acted upon by magnetic force, moving in straight lines and at a high speed away from the cathode; the particles are supposed to have acquired this velocity under the influence of the strong electric field which exists in the neighbourhood of the cathode. We shall call this theory of the cathode rays the "corpuscular theory."

It is from the phenomena discovered by Hertz and Lenard, where the rays seem to penetrate solid substances and even to pass out of the tube, that the ether theory of the cathode rays gets its greatest support. As long as we confine our attention to the phenomena inside the tube, the corpuscular theory seems to give the simplest, and indeed in some cases the only, explanation. Thus take, for example, the curved path described by these rays when moving through a field of magnetic force. The shape of the path follows at once from the corpuscular theory, whereas no attempt has been made to explain it on the ether theory, or to co-ordinate it with any known phenomenon, except for a vague suggestion that it is analogous to the rotation of the plane of polarization in the magnetic field. This suggestion, however, overlooks the fact that the rotation of the plane of polarization is due to the presence of matter in the magnetic field; whereas the bending of the rays is supposed on the ether theory to be independent of the presence of matter. Again, in the case of light waves, which the cathode rays on the ether theory are supposed to resemble, a curved path implies that the waves are travelling at different speeds at different points of their path; but in the case of the cathode rays the path is curved even when the magnetic field is quite uniform, and there is nothing to make the velocity at one part of the path differ from that at another. Thus while on the ether theory the magnetic deflection of the cathode rays is a phenomenon entirely *sui generis* and unconnected with any known phenomenon, it is an immediate consequence of the corpuscular theory.

Again, it would seem hardly possible to get a more direct proof that a stream of negatively electrified particles is an invariable accompaniment of the cathode rays than the experiments described on pp. 161–163, where it was shown that when we catch the cathode rays, we catch at the same time a charge of negative electricity, and that, however we might pull the cathode rays about, we could not dissociate them from the negative charge.

The thermal effects are readily explained on the corpuscular theory by the heating of the substance by the bombardment with the particles, the kinetic energy of the particles being transformed into heat and energy; on the ether theory the thermal effects are easily explained as due to the absorption of radiant energy and its transformation into heat.

The mechanical effects are easily explained on the corpuscular theory by the impact of the particles; on the ether theory they are explained as secondary effects. The thermal effects due to the rays produce differences of temperature at various parts of the tube and at different places on the movable vanes, etc., which may be placed in that tube; then the same cause which produces motion in a radiometer will set the vanes in motion.

The phosphorescence produced by the cathode rays in bodies against which they strike is very easily explained on the "ether theory;" in fact, it seems to have been this phenomenon which first suggested the theory to Goldstein. The phosphorescence produced by the cathode rays is regarded as quite analogous to the phosphorescence produced by ultra-violet light. On the corpuscular theory, too, the impact of the cathode rays will give rise to electro-motive forces, intense while they last, but lasting only for a short time: this condition of rapidly changing electric forces is what on the electro-magnetic theory of light is analogous to the conditions which accompany ultraviolet light. The electro-motive forces produced by the impact of the rays arise in part from the sudden stoppage of an electrified body. This electrified body when in motion acted like an electric current and produced around it a magnetic field; when the body is stopped it no longer acts like an element of current, and does not therefore produce a magnetic field in its neighbourhood; thus the stoppage of the charged particle is accompanied by the destruction of a magnetic field in the neighbourhood of the place against which it strikes. This rapid change in the magnetic field produces an electro-motive force which helps to raise the body to phosphorescence.

We shall now consider the phenomena which occur when the cathode rays fall on very thin plates, as it is these which have, in the minds of many physicists, seemed to tell so strongly in favour of the ether theory as to outweigh all the difficulties which that theory has to surmount in explaining the phenomena which occur inside the tube.

It seems to have been very generally assumed (1) that when cathode rays start from one face of a thin plate after the other side has been struck by such rays, the cathode rays must have travelled as cathode rays right through the plate; (2) That if the rays consisted of very rapidly moving particles they would be incapable of passing through thin films.

Neither of these propositions seems to me to be necessarily true. With regard to the first of these points, we saw (see p. 110) that when the cathode rays fall on a plate of brass 1/16 of an inch thick, a body on the far side of the brass receives a charge of negative electricity; so that one side of this plate becomes a cathode when the other side is struck by cathodic rays. In this case, by the play of ordinary electrical forces and without any passage of the carriers of the electric charges, the cathodic rays have virtually been transmitted from one side of the plate to the other.

We may remark, in passing, that an action of this kind would give a very simple explanation of the fact discovered by Lenard, that the magnetic deflection of the Lenard rays was independent of the nature and pressure of the gas. For if the Lenard rays consist of negatively electrified particles shot off from the window of the tube by the electrical action of the charged particles striking against the window from the inside, since these particles are only in the neighbourhood of the window for a short time, their action on the particles outside will be of the nature of an impulse. The magnitude of this impulse will not depend on the nature of the gas outside the tube: the momentum gained by one of the carriers outside is, however, proportional to this impulse, so that the momentum of the carriers of the Lenard rays will be the same, whatever the pressure or nature of the gas outside the tube. The curvature of the cathode rays in a given magnetic field only depends upon the *momentum* of the carriers, and hence the curvature is independent of the pressure of the gas.

We have seen, however, that the curvature of the rays inside the tube, as well as outside, is independent of the nature of the gas when the electric field is given, and yet the molecules of different gases differ enormously in mass; and in this case we have no reason to suppose that, as in the case of the Lenard rays outside the tube, the circumstances would be such as to make the particles start with the same momentum, whatever the nature of the gas. This result brings us face to face with the question what, on the corpuscular theory, are the carriers of the negative charges? Are they the molecules of gas; or the atoms; or are they matter in some other state of aggregation; and if so, is this state of aggregation coarser or finer than the atomic state?

The experiments made by Lenard on the absorption of the rays outside the tube throw a good deal of light on this point. Lenard found that in air at a pressure of about half an atmosphere the rays could travel about .5 cm. before the intensity of the phosphorescence produced by them on a piece of paper dipped into a solution of pentadekaparatolylketone fell to one half its original value. Now, if we project a molecule of a gas the distance which it will travel before its momentum falls to one half its initial value would be comparable with the mean free path of the molecule, this in air at half an atmosphere pressure is about 2×10^{-5} cm., — a length of quite a different order from the .5 cm. found by Lenard in his experiments. If we suppose that the same molecule does not always carry the charge, but that when a charged molecule, A, strikes an uncharged one, B, the charge passes from A to B, the effect of the obliquity of the collisions would still make the momentum of the molecule carrying the charge fall to one half of the momentum possessed by the projected molecule in the course of a length equal to a not very large multiple of the mean free path. Thus we cannot, in the light of Lenard's experiments, suppose that the dimensions of the carriers of the negative charges are comparable with those of ordinary molecules. If these carriers are not of the same order, the next question is whether they are larger or smaller than ordinary molecules or atoms. We have seen that under certain circumstances a charged particle became the centre of an aggregation whose dimensions must be large compared with those of an ordinary molecule; we might suppose that something of the same kind took place in the cathode rays, and that the long distance they travel before losing half their momentum is due to the great mass of these carriers. This view is not, however, consistent with the fact discovered by Lenard, that the distance traversed by the rays is inversely proportional to the density of the gas. For a large aggregation of molecules would behave like a solid moving through the gas, and the distance it travelled before its momentum fell to a given fraction of its initial value would be inversely proportional to the coefficient of viscosity, and this coefficient is independent of the density of the gas; thus if the carriers were aggregations of a large number of molecules, the cathode rays ought to travel as far in a gas at a high pressure as in a gas at a low one, and this is not the case. Thus, as the carriers are not larger than the molecules, there remains the alternative that the carriers are small compared with ordinary atoms or molecules; and this assumption is consistent, I think, with all we know about the behaviour of these rays. It may appear at first sight a somewhat startling assumption in a state more subdivided than the ordinary atom; but a hypothesis which would involve somewhat similar assumptions — namely, that the so-called elements are compounds of some primordial element — has been put forward from time to time by various chemists. Thus Prout believed that the elements were all made up of the atoms of hydrogen, while Mr. Norman Lockyer has advanced weighty arguments founded on spectroscopic consideration in favour of the composite nature of the so-called elements. With reference to Prout's hypothesis, that if we are to explain the cathode rays as due to the motion of small bodies, these bodies must be very small compared with an atom of hydrogen, so that on this view the primordial element cannot be hydrogen.

Let us trace the consequence of supposing that the atoms of the elements are aggregations of very small particles, all similar to each other. We shall call these small particles corpuscles, so that the atoms of ordinary elements are made up of corpuscles and holes, the holes being predominant. Let us suppose that at the cathode some of the atoms of the gas get split up into these corpuscles, and that these, moving at high velocities and charged with negative electricity, form the cathode rays. The distance these rays would travel before losing a given fraction of their momentum would be proportional to the mean free path of the corpuscles. Now, the things these small corpuscles strike against are other corpuscles and not the molecule considered as a whole; they are supposed to be able to thread their way through the interstices in the molecule. Thus the mean free path of the corpuscles would be inversely proportional to the number of corpuscles in unit volume, and not to the number of molecules. Now, as each of these corpuscles has the same mass, the number of corpuscles in unit volume will be proportional to the mass of unit volume; that is, to the density of the substance, whatever be its chemical nature or physical state. Thus the mean free path of the corpuscles, and therefore the distance the cathode rays travel, will depend merely upon the density of the substance, and not upon its nature or state. This is exactly the result which Lenard proved to be true for his rays.

We see, too, why inside the tube the magnetic deflection is the same, whatever be the nature of the gas; for the carriers of the electricity are the "corpuscles," and these are the same, whatever gas is used. All the carriers may not be reduced to their lowest dimensions. Some may be aggregates of two or more corpuscles, and, being larger, would have a shorter free path and be more easily absorbed and deflected.

I have endeavoured to get some information as to the size of the carriers by the following method: C is a cathode, the rays from which travel through a slit in the cylinder *A*, which is connected with the earth. After passing through this slit, the rays travel through the opening into the inner cylinder, *B*, which is connected with a quadrant electrometer. A little distance behind the slit in *B*, a flat thermo-electric junction is placed far enough behind the slit to allow the rays to enter the cylinder, and sufficiently broad to insure that all the rays which enter the slit strike against the thermo-electric junction. The junction was of iron and copper, and was connected up with a low resistance galvanometer. When the rays struck against the junction it got hot; and from the deflection of the galvanometer the temperature of the junction, and hence the amount of heat communicated to it, can be calculated. The deflection of the electrometer gives the negative charge which passes into the cylinder. Let *N* be the number of corpuscles which enter the cylinder, *e* the charge on each of these corpuscles, *m* its mass, and *v* the velocity. Let *Q* be the charge passing into the cylinder, *W* the mechanical equivalent of the heat given to the thermal junction, ρ the radius of curvature of the cathode rays in a uniform magnetic field of strength, *H*, then we have the following relations between these quantities:—

CATHODE RAYS

$$Ne = Q;$$
$$\tfrac{1}{2} Nmv^2 = W;$$
$$\frac{mv}{e} = H\rho.$$

From these we get,

$$\frac{m}{e} = \tfrac{1}{2} \frac{QH^2\rho^2}{W};$$
$$v = \frac{2W}{QH\rho}, —$$

relations which enable us to determine $m\ |\ e$ and v.

The following are the results of an experiment of this kind. The cathode was slightly curved so as to focus the rays on the slit; the coil was turned on and quickly turned off again. The charge which entered the cylinder raised a capacity of 1.5 microfarads to 16 volts. Hence, —

$$Q = 1.5 \times 10^{-15} \times 16 \times 10^{-8};$$
$$= 2.4 \times 10^{-6}.$$

The impact of the cathode rays on the thermo-electric junction was sufficient to produce in the thermo-electric circuit an electro-motive force of about 52 micro-volts. Taking 16 micro-volts as corresponding to a temperature difference of 1 degree, this means a rise in temperature of the junction of 3.37.

The weight of the iron in the junction $= .021$ gr.
" " " " copper in the junction $= .03$

Hence the water equivalent of the junction $= .0051$. Hence W, the mechanical equivalent of the heat communicated to the junction, is given by the equation, —

$$W = 3.3 \times .005 \times 4.2 \times 10^7$$
$$= 6.3 \times 10^5.$$

The rays of curvature of the cathode rays in a field of 35 units was about 9 cm., hence

$$H\rho = 315.$$

Substituting these values, we get,

$$\frac{m}{e} = \tfrac{1}{2}\frac{2.4 \times 10^{-6} \times \overline{315}^2}{6.3 \times 10^5}$$
$$= 2 \times 10^{-7}, \text{ approximately.}$$
$$v = 1.5 \times 10^9.$$

From these values we get,

$$\tfrac{1}{2}\frac{m v^2}{e} = 2.6 \times 10^{11}.$$

The corpuscle would acquire this energy by a fall through a potential difference of 2,600 volts. In the case of the electrolysis of liquids and of gases, by the ordinary discharge m/e for hydrogen is equal to 10^{-4}, for the corpuscles in the cathode rays m/e is only about 1/500 of this value; thus either the charge carried by the corpuscle must be much greater than that carried by the ion of an electrolyte, or the mass of the corpuscle must be very small compared with that of the ion. The results of this investigation thus support the view that in the corpuscles in the cathode rays we have matter in a finer state of subdivision than the ordinary atom. These small corpuscles may be able to pass through thin solids and thus produce the Lenard rays.

It is worthy of remark that the value we have found for m/e for the corpuscles in the cathode rays is of the same order as that found by Zeemann[1] for the value of this quantity for the part of the sodium atom whose movement produces the D line. Zeemann deduced this from the effect produced by a magnetic field on the period of rotation of sodium light.

[1] Zeemann, Phil. Mag., 43, p. 226, 1897.

www.ingramcontent.com/pod-product-compliance
Lightning Source LLC
Chambersburg PA
CBHW030109170426
43198CB00009B/554